With PTARMIGAN & TUNDRA WOLVES

CY HAMPSON

Orca Book Publishers

Canadian Cataloguing in Publication Data
Hampson, Cy, 1914 -
  With ptarmigan and tundra wolves

  ISBN 0-920501-59-1

  1. Animal behavior – Arctic regions.
I. Title.
QL751.H35 1991   591.51'09719   C91-091113-4

Cover design by Susan Fergusson
Photography by the author

Printed and bound in Hong Kong

Orca Book Publishers
P.O. Box 5626, Station B
Victoria, BC
Canada V8R 6S4

To my wife, Mary, who works enthusiastically with me in far-flung corners of the world; to the wildlife of the Arctic whose strategies for survival continue to challenge our thinking; to those resolute researchers who would add yet another fragment of knowledge to life in the fragile Arctic.

Acknowledgements

I am deeply indebted to Dr. T.W. Barry, Canadian Wildlife Service, and his talented wife Dr. Pat Barry who first introduced my wife and me to the Anderson River delta in 1963. They were also most helpful when Mary and I returned years later to carry out a study on the "Reproductive Behaviour of the Willow Ptarmigan," sponsored by the Boreal Institute. I also wish to express my appreciation for the assistance of the National Museum of Natural Sciences, Ottawa, and the Polar Continental Shelf Project both of which sponsored the expedition to Bathurst Island.

S.D. MacDonald, curator of Vertebrate Ethology, National Museum of Natural Sciences, Ottawa, led the Bathurst Island Expedition with his customary thoroughness. If there is a key to success in such an enterprise, it lies in thoughtful organization and the careful consideration of a thousand relevant details.

While the National Museum was primarily interested in research into animal behaviour, under the leadership of S.D. MacDonald, the Polar Continental Shelf organization was concerned with accumulating broader, base line data pertaining to the Canadian Arctic.

In addition, I should like to thank Gray Campbell for reading the manuscript.

In the text, two or three names have been changed in deference to the individuals concerned.

Contents

Foreword   ix

Update   x

Inuvik   1

The Anderson River   9

Grizzlies   16

The Key   27

The Impossible, Possible?   36

Past and Present   48

Visitors   62

New Beginnings   71

To Bathurst Island   78

The Arctic Hare   87

The Elusive Sanderling   99

Tundra Wolves   112

Neighbours   119

Foreword

This book is primarily a work dealing with the intriguing behaviourial patterns and strategies which have evolved in a number of arctic animals, enabling them to survive in the exacting and often formidable land far beyond the Arctic Circle.

The regions represented are the Anderson River Delta in the Western Arctic and Bathurst Island, lying in the Central Arctic less than 15 degrees from the North Pole. This was the last of the High Arctic islands still to be explored.

While the book is based upon the first-hand experiences of the author, much information was gleaned from those with whom he worked. In addition, the work endeavours to capture the indomitable spirit and character of those who regularly leave comfortable southern homes in order to respond to the challenge of research in these remote regions. Their drive has always been the often elusive and painstaking search for truth, the attempt to untangle some of the intricate threads in the fabric of northern life and expand our knowledge of those animals which have been able to adapt successfully to life in this fragile and often inhospitable land.

Update

The exercises related here were not without significance. Dr. Dave Parmelee produced a superb paper on the "Breeding Behaviour of The Sanderling in The Canadian Arctic" and S.D. MacDonald an equally fine one on "The Breeding Behaviour of the Rock Ptarmigan," both published in *The Living Bird*, Cornell Laboratory of Ornithology. David Gray continued his excellent work on the behaviour of muskoxen in the High Arctic and thereby earned his doctorate from the University of Alberta. David's book on muskox behaviour has been published and he is currently working for the National Museum of Natural Sciences, Ottawa, while Phil Taylor has accepted a position with the Canadian Wildlife Service in Saskatoon, Saskatchewan.

Mary and I added to the library of the Boreal Institute a carefully documented paper on "The Breeding Behaviour and Reproductive Success of the Willow Ptarmigan on the Anderson Delta." The author also produced a half-hour educational film based on the Anderson River delta and another of equal length, based on the expedition to Bathurst Island.

In addition, a good deal of information was gained in connection with tundra swan and tundra wolf behaviour, as well as extensive insight into the life of the Arctic hare, accurate data on the incidence of flora and fauna of the areas and prevailing weather conditions.

The cabin on the delta of the Anderson River continues to serve as a valuable research station in the Western Arctic. Due to sustained effort on the part of S.D. MacDonald over a period of eighteen years, the Government of Canada, on September 20, 1986, created the Polar Bear Pass National Wildlife Area. The site of our original base camp was the nucleus of this reserve.

Inuvik

Mary and I were gazing down, spellbound, through the small window of the Twin Beach aircraft upon the great sprawling watercourses of the delta of the mighty Mackenzie River, 150 miles inside the Arctic Circle. The myriad of still icebound, twisted and contorted ribbon-like channels were fringed with dark vegetation, forming a veritable maze in sharp relief through which it appeared impossible for a mere human to thread his way.

"I wouldn't care to be lost down there," Mary ventured.

"Nor I. I'd want a strong sled and a reliable team of experienced huskies that knew their way, for sure."

We had stopped briefly at Fort Smith, Hay River, Norman Wells and Fort Simpson on our way from Edmonton, Alberta, and were bound again for the delta of the Anderson River on the Arctic coast, northeast of Inuvik. The time was early May. For years, the very sound of such terms as Inuvik, Aklavik, Tuktoyaktuk, Resolute Bay, Coppermine, Sachs Harbour, Winter Harbour and Perry River had stirred my imagination. So, too, had accounts of the expeditions of Franklin, Ross, Richardson, Henry Hudson, Rasmussen, Freuchen, Peary, Stefansson and a dozen others.

Nor had my level of interest been lowered by long chats with Dr. J. Dewey Soper, who had traversed much of this country by dog team. Our first trip to the Arctic coast had fixed the North in our blood. We were now in need of another transfusion.

But since 1963, we had been reading widely and carefully trying to digest our former lengthy notes with the hope of shedding some light on how wildlife manages to survive on our last frontier, far from more familiar regions further south. The sum total of our experiences only served to reinforce the basic premise that the Arctic, with its incredibly thin layer of soil above the permafrost comprises an environment that is fragile in the extreme. And coupled with this fragility is a climate characterized by long, hard, dark winters and very short growing seasons which sharply limit the productivity of that environment.

Breeding seasons, especially those of migrants, must of necessity be abrupt and brief. Nevertheless, many animals and plants do survive and even flourish in these seemingly inhospitable, harsh surroundings.

Admittedly, while the majority of the forms of animal life is comprised of visitors from farther south, which return annually to breed here, some species are year-round residents. The compelling challenge was to discover what we could about the means by which the forms we were to encounter firsthand are able to survive under these exacting conditions. From having worked in East Africa for a short period of time, we realized that interrelationships here, involving a great many fewer species of plants and animals, should be less complex and more readily comprehended. Yet in these situations, there always remains a host of intriguing but unanswered questions. In fact, such studies and observations invariably tend to raise more questions than they answer.

Shortly, we began losing altitude rapidly and at last coasted in, landing smoothly on the dark airstrip of Inuvik, the "Place of Man." Jerry, a member of the staff of the Canadian Wildlife Service, was waiting with a half-ton truck, and we transferred our personal belongings into it and drove into the northern village, dotted somewhat sparsely with stunted, slender black spruce, aspen and balsam poplar. The trees were there, well within the Arctic Circle, the vagaries of the delta climate together with the soil composition and availability of essential water having given them a foothold. We were later to learn that human residents prized their trees. In one instance, a bulldozer operator had been peremptorily fired for accidentally uprooting a single tree!

Inuvik was unlike any other settlement we had seen. Brightly painted residences, schools, a restaurant and Hudson's Bay store in the "serviced area" were perched upon stout piles, as one might expect to see the huts of natives in a tropical region of mangrove swamp. Curious, long, box-like galvanized containers connect the buildings with a central steam-heating plant. These containers, or "utilidors" are insulated and carry the hot steam lines along with the water and sewage pipes. None of the contents are frozen and the residents enjoy reliable sources of heat and water as well as an efficient method of sewage disposal.

Inuvik is a village constructed on permafrost. Were it necessary to install such utilities below the frost line in the conventional southern manner, the engineers here would have had to excavate to unfathomed depths in order to install their pipes. The extensive air spaces under the buildings provide a sort of breezeway allowing the free circulation of air which removes radiating heat from the structures and prevents the thawing of the permafrost below. Otherwise, large, well-heated buildings would slowly sink below the surface.

In the "unserviced areas," the small dwellings are insulated from the permafrost by very thick gravel pads. Here, the residents heat by stove or space heater, carry their water by hand, usually, and have their garbage removed by truck.

Inuvik is a young settlement. It is the site of the "New Aklavik," and only by the summer of 1956 had it begun to take on the hesitant profiles of a town. Many hundreds of piles had to be floated to this site on the east arm of the Mackenzie and the upper layer of permafrost partially melted with steam hoses so that the piles could be driven home with massive pile drivers. During the following winter, they froze in firmly and carpenters could proceed with the superstructure.

Inuvik was designed originally to supersede the village of Aklavik, the "Place of the Brown Bear," thirty-five miles distant on the west side of the delta. Standing at the end of the long river route of the Athabasca, Slave and Mackenzie Rivers, Inuvik was to be the new administrative centre for an enormous region of some 350,000 square miles. This would take in the lower Mackenzie Valley, and

The hawk owl breeds north to the tree limit in Canada and Alaska.

the part of the Arctic coastline and nearby islands extending from Herschel Island, near the Alaska-Yukon boundary, to Spence Bay, located in the Central Arctic a thousand miles to the east.

The abandonment of Aklavik as the administrative centre for this region was dictated by the fact that this village had nowhere to expand and provide the facilities required by a rapidly developing area. A veritable prisoner of the river, it was hemmed in against the Mackenzie Channel by extensive lakes and marshes. In addition, Aklavik consisted of a low-lying site, dangerously plagued and threatened by annual spring floods at the time of breakup. The river was gradually eroding the fringes of the village. Another critical problem lay in the unstable nature of the soil, the gooey, sucking, heaving mud of such consistency that it could not support satisfactory foundations for modern buildings. There was some reason for dubbing Aklavik "the mudtropolis of the north."

We were anxious to leave for the Anderson delta, some 150 air miles to the northeast, but adverse weather intervened. A dense grey fog moved in on a nor'wester from the icepack off shore, forming an impenetrable curtain for the pilots. We were fogbound.

But this gave us an opportunity to look around the village, to wander along the narrow roads still bordered with banks of greying, wasting snow which had barely begun to thaw. The centrally located Sir Alexander Mackenzie residential school was impressive and attended largely by youngsters from outlying native settlements, who were flown into Inuvik for most of the year. Brief holidays in summer were spent with their families back in their isolated, outlying settlements.

Ravens are permanent residents in much of Canada, breeding as far north as many of the High Arctic islands.

The most unusual Catholic "Igloo Church", looming up in the fog, immense and ethereal, was an extraordinary but thoroughly appropriate sight.

The "native quarter" consisted of a series of low log cabins, each with its quota of sled dogs out behind. These people were mostly hunters and trappers, harvesting the rich resources of the delta country, such as muskrat, mink, marten, weasel and fox. Coastal regions offered seal, white whales and the incomparably flavoured arctic char, probably the finest table fish in the world. A local trapper mentioned that he had taken several northern flying squirrels in his marten sets. These interesting, noncommercial squirrels were certainly living at the northernmost limits of their range in Canada.

We watched the native peoples in their parkas as they sawed wood, fed their dogs, repaired sleds and carried water in large buckets. One day, we were standing under a small black spruce watching the smoke curling lazily from the chimneys of the cabins when Mary suddenly turned to me.

"The ravens here are a different species from those we get at home, aren't they?"

"I don't think so. I know of only one species in Canada. What makes you think they are different?"

"Well, look at them! See the one over there on that tree?" she said, pointing with her mittened hand. I followed her pointing hand and immediately spotted the ebony bird, perched on a drooping branch near the top of the tree.

"Just an ordinary raven," I replied.

"Ordinary, my foot! Look at the tail."

The pigeon hawk, or merlin, breeds widely across Canada as far north as the Mackenzie Delta and Alaska.

The raven didn't have one. Certainly, no more than a stump, and a mighty short stump at that.

"There are two more over there," pressed Mary. "And they are just the same."

At that moment, still another bird left its perch and circled above us. The usual well-developed, rounded tail was missing. In a moment a figure left one of the cabins, heading in our direction with an empty bucket in either hand. When he came abreast of us, I raised my voice. "Excuse me, please, but I am curious about your ravens. Don't they ever have tails?"

"Yes, for awhile," he answered, his round face wrinkling into a gentle smile beneath his parka hood. "You are from Outside, no?"

"Yes, but do you know why none of these birds have tails?" I swept my arm in an arc in front of me, indicating the ravens in the trees to our right.

"Yes." His eyes were twinkling now.

"Could you tell me then, please? I've never before seen an 'awhile tail' on any bird."

"Yes, but you like to study birds yourself, no?" He pointed at the field glasses slung around my neck.

"Yes, we do."

"Then I would not like to spoil your fun. You stand here and watch and you will learn." And he turned away with the same gentle smile raising his full lips slightly at the corners.

"He wouldn't tell us," I said to Mary. "So I guess that we'll just have to watch. But I'll be an anthropoid's uncle if I know what to look for."

The air was decidedly nippy, and after waiting for nearly an hour, we decided that the whole thing was a great joke. We were about to leave when the dogs tethered behind one of the cabins set up a great clamour, barking and howling. On the instant, the door opened and a rotund, parka-clad figure emerged carrying a pan, held horizontally. He started towards the dogs, at which they increased their volume of yelping and began straining upward at the ends of their chains. The man stopped briefly and tossed the contents of the pan in the direction of the leaping, hungry huskies.

But before the food had reached the snow, a big raven hurtled from a perch above and snatched a beakfull from midair. A snarling husky had lunged forward at the same instant. His jaws closed on the very north end of a raven heading south and the dog held the bird's last three tail feathers between its teeth. The Eskimo had not spoiled the fun of our discovering the unknown.

There was still a decided chill in the air as we headed back towards the Anglican hostel having heard that these public-minded people always had a pot of coffee on the stove and welcomed visitors. We were just passing the Igloo Church when a small group of people emerged from the door and began to move slowly towards a light truck and three cars parked out front. Four men at the front were carrying a small wooden box, two on either side.

The procession, if that is what it was, proceeded with heads down and stood quietly while the box was loaded carefully into the truck. We watched them drive away in single file, wondering where they would go with so very few roads in evidence. We had seen several cars at the airport, a short distance from the village, when our plane had landed. We had hardly started on again when we noticed a woman waving her arms from the step of a porch and calling loudly.

"Come on over, Mary! We've been expecting you."

"She must mean us," Mary said. "But she can't. We don't know anybody up here but Jerry and his family and the fellow from Aklavik Flying Service who is going to fly us out to the Anderson again." But we hurried forward to the woman who was still beckoning.

"Don't you know me, Mary?" she asked.

"Gwen, it's you!" cried Mary. "I didn't know that you were in Inuvik!" They threw their arms around one another and chortled in glee.

"Then you didn't know that Henry had left the station in Whitehorse? We love it here. Of course, he's still doing the weather for the Department of Transport. But this must be Cy. We've been looking forward to meeting you." And she gave me a bear hug and a kiss, her eyes alive with excitement.

"Did you say 'Gwen', Mary?"

"Yes. Gwen and Henry Mody. Gwen and I were close friends in the Edmonton Y."

"Look, the coffee pot is on and I've just taken new bread out of the oven. Quick, nobody wants cold coffee."

When we went inside, the aroma of coffee and fresh bread filled the kitchen to overflowing. In a trice, Gwen had poured coffee and cut three generous slices from one of the loaves on the counter by the stove. We slipped off our parkas and sat down.

"Peanut butter or strawberry jam? That's about all that I can offer you until

The northern flying squirrel is a permanent resident in most of wooded Canada and Alaska, including the Mackenzie Delta.

the barge comes in this summer. Air freight is very high so we try to order ahead as much as we can so that it comes in by way of Northern Transportation down the Mackenzie. I thought that I had ordered lots of everything, but the neighbours have been borrowing a lot more than I had bargained for. But they're not alone. I borrow, too. We all do, towards spring." Gwen's warm cheerful smile was infectious.

"Jam sounds great, Gwen. Now, how did you know that we were here? And how did you know Cy's name?" queried Mary. "I haven't seen you since we were married. Lost track of you altogether."

"You mean that you haven't heard of the moccasin telegraph? Lots of people up here wear moccasins, you know." Gwen was grinning broadly. I figured that she'd keep Mary guessing awhile.

The girls continued swapping information and recollections over three cups of coffee and half a loaf of delicious bread. All too soon, we felt that it was time to leave.

"Gail and Eric are at school and Henry will be home about supper time. He'd love to meet you, so come on over when you're free. So you're going out to the Anderson to photograph birds and flowers? And work on ptarmigan?"

"Gwen Mody, you're incredible!" Mary smiled. "Oh, by the way, was there a funeral in Inuvik today? We saw what we thought was a small coffin being carried out to a truck over by the church an hour or so ago."

"Yes, the small son of one of the airport workers died of pneumonia three days ago. An awfully nice little boy, just ten years old. I saw them taking the steam hose up to the cemetery yesterday morning."

"The steam hose?" Mary asked.

"Yes, they have to thaw out the permafrost in order to excavate the grave. The same way they do with the pilings for our houses."

The adaptable coyote has extended its range into the Middle Arctic.

Mary shivered at the thought, and then said, "I somehow don't like the idea of entombing people in ice for hundreds or thousands of years. Especially young girls and boys. In the South, you know that they will soon return to the soil. Do people here ever think of their friends down there, frozen almost as if they were alive?" Later, Mary expressed the same sentiments when the poor fellows from the Franklin Expedition were found beneath cairns, buried in graves of ice.

"You do what you have to do, don't you," Gwen said quietly, supportingly. "See you for dinner, then. Today, around five?"

"Thanks, Gwen."

The Anderson

Only tenuous wisps of the thick fog remained a few days later when Jerry assisted us in loading our equipment onto two Cessna 185's resting on the snow-covered ice on the river. The sun, shining brightly in the pale sky, turned the smooth, packed runway into a sheet of gleaming satin bordered on the far side with stunted black spruce. There was a delicious crispness in the air which monitored our breathing in miniature clouds of vapour rising from beneath our parka hoods. A dark raven, sans tail, circled lazily above, croaking hoarsely at intervals.

"Are you warm enough?" I asked Mary.

"You bet! Woollen underwear, corduroy pants and this heavy, lined parka would be hard to beat. And I still prefer leather mittens with woollen liners."

"Jerry has disappeared. Any idea where he has gone?"

"He said that he had forgotten something."

Jerry appeared again shortly, carrying a rifle and a box of cartridges. "Here, you'd better take the grizzly gun, Cy. But you won't likely need it. I've only used it once out there."

"Thanks."

Jerry continued, "By the way, I accidentally dropped it in the ocean last summer and just haven't had time to work it over. It may need a bit of attention."

Jerry would accompany us on this trip to the Anderson as he wanted to check out the cabin, the freighter canoe, the fuel for the outboard motor, the radio, and other details in connection with our stay of three months or more. On walking out to the loaded planes, I noticed the stout, efficient-looking metal skis with which they had been equipped for the long winter season. The snow underfoot had definitely lost the sharp crunch characteristic of very low temperatures and it seemed that spring was surely upon us. Jerry introduced us to the two pilots and after they had climbed aboard, he turned to me.

"Your man is new to the Arctic," he said, indicating our pilot with an upward jerk of his thumb. "But I'm sure he'll be all right. His recommendations are good."

The ski-equipped planes glided smoothly upriver and upwind, accelerated swiftly and in a moment were airborne. We were off at last. The planes circled the village, climbing steadily. From the vantage of height, Inuvik looked trim and orderly with cabins and residences in neat rows. Schools, the hospital, power plant, skating rink and Igloo Church could all be easily spotted. From the air one could more fully appreciate the planning that had gone into choosing the site and developing it. The proximity of the village to the large navigable river and to the high ground occupied by the airport were very significant advantages.

The pilots turned the noses of the planes eastward. We were flying over a snow-clad landscape with twisting streams, bordered with scant, dark vegetation, many of the larger loops almost completely doubling back on themselves. Somewhat rounded white expanses, free of any vestige of plant growth, were tundra lakes, very numerous and greatly varied in size. I was captivated by the series of strange geometrical shapes that were clearly visible at several points along the margins of the larger lakes and leaned forward to peer at them more carefully. They were polygons, frost polygons, created by upheaval during deep freezing. We would look at some of them more closely from ground level.

In about an hour and a half, we were over a larger stream and turned northward following it. Could this be the Anderson again? I couldn't be sure. On our first glimpse of it from the air, years earlier, it had been shrouded in fog.

In a few minutes, the main stream below broke into several smaller diverging channels emanating like the sun-bleached roots of a wind-fallen tree from the main trunk, and we were over a delta. The areas surrounded by curving loops of adjacent small frozen water courses indicated islands in the broad delta.

"There!" cried Mary, pointing vigorously with her mittened hand. She had spotted a tiny white cabin, just above the river and clearly outlined against a vast expanse of dark brown willow tundra. The roof was bright red in colour and a second smaller shed-like building stood beside it. This was to be our home for the next three or four months. A home totally isolated from neighbours by 150 air miles of open, trackless, stream-crossed and lake-dotted tundra. We were eager to tenant it.

As we began to lose altitude in preparation for landing, it could be seen that a strong ground wind was whipping skeins of snow southward like gale-borne smoke over the frozen surface of the main channel below the cabin. Nevertheless, our pilot swung his plane around to the north and began to come in for a landing with the wind in his back rather than against it.

"You aren't going to try to land 'er down wind?" I shouted, straining to make myself heard above the roar of the motor and the high-pitched whoosh of wind through the struts.

"Damn right," barked the new pilot from Outside. "What do they call those white whales they get up here?"

"Belugas."

"Well, it's as smooth as a beluga's belly down there. Waste of time circling when you don't need to."

"You're dead wrong, Buster. It only looks smooth. No sun. No contrast. The wind whips that stuff into cross drifts as hard as concrete. Better go around again and come in against the wind. And slower. It'd be a lot safer."

"Look, you can't get a pilot's license with poor eyesight. Think I can't see?" he growled. "Anyway, who's flying this banana crate? I've got two more trips to make after this one, so hang on!"

"Looks as though a damn fool is flying 'er," I flung back, seething with anger, but he wasn't listening. The Cessna dropped steadily, the wind square in its tail. She was headed for the broad area in front of the cabin, now nearly lost

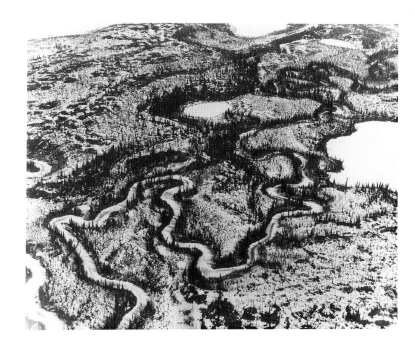

*Tortuous frozen
streams, east
of Inuvik.*

behind an enormous drift of snow along the bank. The pilot accelerated still more as the plane levelled out about a hundred feet above the snow. I realized that he couldn't brake in the running wind and still keep her nose steady. Ground level rose as the short dark willows on the near bank stampeded by.

"Hang onto anything, Mary!"

I tried in vain to relax as the skis slammed into a rock-hard drift with a loud, wrenching "crack!" and then recoiled into the air again.

"Don't go over on your back, please, baby!" I had instant visions of spilled gasoline, a hot engine and billowing, consuming flames.

We bounced at least twice more before the loaded plane came to a shuddering, curving halt on a single intact ski. When we climbed out, we found the other ski a mass of twisted metal.

"We are very lucky," I said, striving to contain my anger, mixed with a sense of relief. "And you are an absolute, blasted fool."

The pilot in the second plane was circling to come in against the wind. He glided in slowly and bumped over the hard, rounded contours without much trouble. His door burst open and he rushed to the first plane, arms waving.

"And where the hell did you learn to fly?" he snarled. "Geez, I thought you said you were experienced!"

"I . . . I didn't know it was rough. It looked smooth from up there."

"Rough or smooth, you don't know a damn thing about flying! Did they really teach you to land downwind on a surface you'd never seen before?" He bent over to check the undercarriage and immediately spotted the useless ski. "You silly, tinhorn sonova so'n'so, you've smashed the undercarriage. And two more trips to make today. Nobody can run a business this way."

I . . . I'll work overtime and . . . and you can take it out of my wages," the pilot said lamely. His face had gone the colour of the blowing snow.

"Like hell you'll work overtime. Not for me, you won't. You could have flipped 'er! Passengers aren't safe with you. You'll stay right here until I bring another ski from Inuvik. You'll replace the broken one and then fly 'er back. The second you touch down on the river, your time's up. I'll be there when you get in, you jackass, and you'd better check the wind this time before you set her down! Hey, did you think of checking with Cy about the surface of these frozen rivers up here? He's been on them before."

"I never thought of that," he said, almost under his breath.

I had begun to feel that the young pilot had had a pretty tough break so early in his northern career, but his answer changed all that. He'd be a definite liability in the North. We all set to and in no time all of our gear, including a snow toboggan, had been unloaded onto the ice.

"You'd better start by getting that wrecked ski off and if you have any time left, give these folks a hand with their gear."

And with that the second pilot, who was evidently in charge of the operation, climbed into his plane and waved. The engine coughed, caught, steadied into a roar and the unloaded plane was climbing steeply into the wind. Jerry, who had been listening quietly to the whole altercation, began loading the toboggan and we started pulling the first load up the steep bank to the cabin. This involved making our way up a forty-foot snowdrift that had begun to thaw in spots, allowing us to sink to our hips in places. The great drifts were not the result of heavy precipitation, as the general area tends to be a northern semi-desert. However, strong, unimpeded arctic winds whip whatever snow there is across the tundra for enormous distances, depositing it in depressions and on the lee sides of steep hills and the banks of streams and rivers.

Much of the cabin door was barred with a huge drift of snow and the roof splashed with whitewash from perching ptarmigan. We noticed that the door and windows had been covered with heavy shutters through which a great many long nails had been driven, their sharp ends protruding well beyond the surface of the wood. Jerry explained that this was "raider protection," that during previous seasons the cabin had been raided by wolverine and grizzly bears which had played havoc with the place in their efforts to get at unused supplies stored in the attic at the end of each summer.

"If you two would like to continue hauling up your gear, I'll just get a shovel from the shed, clear away this stuff, set up the radio, show you the maps of the delta and tell you a bit about some of the narrow, rather difficult, delta channels that can pose a problem or two until you get to know them."

"Great!" Mary and I chimed in. Three hours or so later, we had just about finished when the drone of an approaching plane announced the return of the Cessna from Inuvik. Jerry had cleared away the snow, removed the shutters, started a cosy fire in the oil drum stove and was now working on the radio.

"How would you like to put some water on the gasoline stove, Mary? It's time we all had a bit of lunch. Just melt some snow in that big old kettle, there. Snow will be your water supply until breakup."

The isolated cabin on the edge of the Anderson Delta had been originally built for Eskimo reindeer herders.

"You bet," Mary agreed enthusiastically. "By the way, I've forgotten the name of the pilot who is coming in."

"We call him Mac. He's a good sort."

Mary had the kettle boiling by the time the two pilots appeared. "What will it be, fellows?"

"Tea," said Mac. "I guess I'm half-Eskimo by now." And tea it was all round. Tea, crackers and cheese. It tasted great after all the exertion. Most of us had two or three cups. Then Mac turned to Jerry.

"We'll be about half an hour getting the new ski on, Jerry. Have you got a few things you'd like to talk over with Mary and Cy before you leave? We won't likely be out again."

"You bet I have."

Jerry went over to a large map pinned on the east wall and pointed out the general layout of the delta, the streams that had narrow channels, especially at low tide, the importance of pulling the canoe well up when working the islands in the tidal waters of the delta and the preferred breeding areas for various species. Then he showed us their phenology charts of the area, which gave accumulated data on sightings, nesting, hatching, banding of waterfowl, etc.

Finally, he showed us how to operate the radio, explaining that it was not always possible to get through with messages. If the opportunity arose, he'd come out for a bit of a check sometime during the summer, but he couldn't promise. He had a lot of irons in the fire.

"Now, is there anything else?" he grinned.

"I don't think so, Jerry. I can't tell you how much we appreciate all of your help and kindness. We'll never be able to thank you adequately."

"Have a fine season. Get some great shots, and I want to hear all about your ptarmigan study when the summer is over," he said, extending his hand. We shook hands all round. He slipped on his parka and vanished over the lip of the bank, heading for the waiting planes.

"What a fine fellow," Mary said. "We owe him so much."

"Right you are. Now I think that I should saw up a bit more driftwood for the stove and then we'll pack things away."

"Let's have a look at the big map first," countered Mary. "I like to know where I am."

We moved to the two maps pinned to the wall, one showing details of the delta and the other, a general map, covering much of the region. The Anderson rises well south of the tree line in the vicinity of Great Bear Lake, the southern portion of its course flowing through the taiga, or boreal forest, where it would pick up the spruce, larch and birch driftwood which we burned in our stove. The river then enters the Arctic Ocean via Wood Bay, Liverpool Bay and the Beaufort Sea, which is, in reality, part of the great circumpolar ocean. The other map showed the many watercourses and islands of the Inner, Middle and Outer deltas, together with the names which Jerry and his associates had given them over the years during which they had been studying and banding waterfowl in the area.

"Fox Den Island, Study Area, Boat Island, Bluff Island, Moose Track Lake, Triangle, Little Fish Lake, Poosticks Creek, North Bluffs, South Bluffs, Ptarmigan Meadows, Bipsi Lake, Hunt's End Creek," Mary mused aloud, looking closely at the map. "It would be interesting to learn how they all got their names."

"We'll just have to ask Jerry when we get the chance. Hey, how about another cup of tea? Isn't this great? No telephone, no buckets of tragedy on a T.V. or radio, no screaming headlines in the newspaper. But a great chance to sort out a few of the adaptations of animals and plants which enable them to survive here. And I'd like to try to document as much as possible in our notes and on film."

We had tea with crackers and cheese again. Then I began sorting and stacking cartons against an outside wall and under a waist-high work bench, making sure that our scribbled labels showed "flour, corn meal, porridge, dried milk and vegetables, dried fruit, honey, syrup, sugar, some freeze-dried stuff, tea, coffee, the odd cake mix, cheese, peanut butter, canned meat, pancake flour, dried soup, chocolate, nuts, salt and pepper, and two cartons of bread." We would keep the bread at least partially frozen in a barrel in the shed until we could rig up some sort of refrigerator in the permafrost, if that were possible.

Mary had decided to celebrate by trying a couple of pans of corn bread in the drum oven on the stovepipe. This contraption, the drum oven, was a great idea so long as the heat was controlled with the dampers on the stove and pipe. It consisted merely of a double-walled stovepipe with a fair space between the walls and a sort of balloon-like swelling in the pipe. The smoke and heat rose between the layers of pipe and around the "balloon" which contained a horizontal shelf and hinged door on the side. It could bake two loaves of bread, side by side, at one time.

During the winter, snowdrifts accumulated to great depths along the lee side of the river.

"We didn't forget to pack the first-aid kit, did we?" Mary interrupted.

"No, but we must be as careful as possible out here. We just can't afford accidents, you know. We are a long way from help."

"Did Jerry ever tell you when they built this snug little cabin in such a remote area?"

"The Wildlife Service didn't build it. They bought it. It was originally built for two Eskimo families who were to look after the reindeer herd which had been brought over from Europe. But they never used it."

"Never used it? Why not?"

"There were thirteen Eskimos in the two families. They were coming over the Beaufort Sea, north of here, when their boat ran into a violent storm. The boat capsized and they were all drowned. A little later, the reindeer herd was moved up to an area south of Tuktoyaktuk and it's managed there, now. The cabin became available and the Wildlife Service bought it for a research station."

"That must have been a terrible catastrophe for the Eskimo village from which they came. They are such a close, kindly, feeling people, and so helpful to others. It must have been devastating to lose so many friends and relatives all at once! Heartbreaking."

"Yes, but you can be certain that these superb people will have rallied as best they could. It has been this very closeness, together with their ingenuity, which has enabled them to survive against such incredible odds in this environment of extremes."

We had our tea with delicious fresh corn bread and honey and rolled into our bunk beds, tired, but excited about what the next day might bring.

Grizzlies

After a quick breakfast, we donned heavy woollen socks, high rubber boots, parkas and warm mittens and hurried outside to see what wildlife was present on this May 15. The first sounds that assailed our ears were the very exciting, much varied calls of willow ptarmigan engaged in establishing their territories on the willow tundra surrounding the cabin. Within a couple of minutes, we had spotted no fewer than two dozen brilliant males perched upon the highest points available in the immediate terrain, which sloped gently, both north and south, from frozen Little Fish Lake which lay about a half-mile north of the cabin. They were perched conspicuously upon any slight elevation, such as a clump of slightly higher willow, a hummock, the odd discarded gasoline drum or the ends of grey logs protruding, strangely, above the snow at several points in the general vicinity of the cabin. One bird occupied an end of the cabin roof.

At this time, we saw mostly males with their gleaming white bodies, reddish brown heads and necks, scarlet combs and black outer tail feathers. They had lost the almost complete white colouration of winter and assumed the courtship garb. The few females that we spotted were still in the white plumage of winter as they made their way unobtrusively over the surface of the snow among the willows, clearly taking advantage of the partial cover afforded by the vegetation. Suddenly a male would burst high into the air in a swift, towering flight with wings stiffened broadly at the summit. Then a dip and down he came on fluttering wings, at the same time emitting a crescendo of rapid, loud "ca-ca-ca-ca" notes which changed to an explosive "ta-bacca! ta-bacca! ta-bacca! ta-bacca!" given on landing. The sheer volume of sound was almost unbelievable.

It was clear that the males had already begun to establish their territories and were bent upon proclaiming these to potential rivals. At the same time, they were undoubtedly bent on attracting mates, as indicated by both their brightly contrasting colouration and conspicuous behaviour patterns. In short, the males were involved in a very dynamic contest. They had returned to this favourable breeding area from various points farther south and here they were vying for the essential elements necessary for procreating their kind. Choices would be made which in the long term would prove to be in the interests of the survival of the species. Their sheer numbers were indicative of success, at least in this area.

We decided to hike to a small rounded area dubbed Kettle Hole Point on the map. It jutted out into Wood Bay along the coast about three miles distant.

The air was fresh and balmy at the outset, but the snow underfoot had lost its firmness under the urge of the sun's rays and the going was difficult as we kept breaking through the upper crust to our boot-tops. A series of high-pitched calls out of the south drew our attention and we turned in that direction. A long, wavering line of white bodies with black wingtips was approaching steadily and almost on the deck.

"Snow geese! Wavies!" Mary cried.

"Right! And there are more in the distance behind them."

The birds were pumping rhythmically but seemed a bit laboured, as though tired. They were a superb sight against the pale sky as flock after flock swept over, the sun full on them. We knew that they had been moving northward with the advance of spring temperatures, but there was some evidence that the last long leg of their flight to the breeding ground was almost nonstop. Jerry had told us that the geese nesting on the delta tended to be very punctual on their arrival dates, regardless of the weather and snow conditions on the surrounding tundra. There were certainly no bare spots at the moment which could serve as nesting sites. But the breeding season was short and there was no time to waste.

The beach at Wood Bay consisted mostly of golden washed and rounded pebbles glistening in the sunlight. The shoreline curved away into the distance, at length meeting the low horizon. Here there was no bank along the upper shore to catch and hold the blowing snows of winter, and the surface of rounded pebbles had been swept clean. Two pairs of magnificent tundra swans flew over a few yards above our heads, great wings creaking, long necks outstretched, large black lobed feet bringing up the rear.

"If we could only photograph them at the nest," Mary said excitedly.

"One never knows, but I've been told that they are very wary at nesting time. 'Impossible to photograph at the nest', was the way an experienced northerner put it to me."

On the return trip, we saw many more ptarmigan, four plump snowy owls, three of them dark females, and a total of twenty-three long, slim gyrfalcons, all grey phase. We had been told that they breed on the Upper Anderson in trees, which, at the time, seemed entirely extraordinary as we had always associated gyrfalcons with rocky cliffs. However, the tree-nesting behaviour had been adequately documented by Frank Beebe and Richard Fyfe. It looked already as though we should have plenty of species to observe.

On arriving back at the cabin, we were startled to find a series of large tracks of two animals which had come up from the river, circled the cabin several times and then made off in a northward direction.

"Grizzly bears!" The last thing we needed around camp for the season!

One set of tracks was larger than the other, and we assumed that the imprints were those of a pair, probably a large male and a smaller mate. The impressions of the very long claws were anything but reassuring. A search of the area in every direction with field glasses failed to turn up the bears.

"We are very fortunate that they did not raid the place while we were away,"

I said. "All of our supplies are in there, together with the rifle and the radio, which is our only way of contacting anybody."

"It was a grizzly that made all of those gouges in the ceiling of the cabin, wasn't it?"

"That's what Jerry said. And he must have been a good size. Those walls are eight feet high. Maybe we won't see them again, but I'm going to load up the rifle just in case."

Loading up the rifle was no easy matter. I'd forgotten that Jerry had accidentally dropped it in the ocean the previous year. The cartridge clip could be removed only after much persuasion, and the bolt in the breech had been frozen with briny rust. After several hours, working with generous portions of thin oil and fine emery cloth, the breech could be opened, and all parts oiled and cleaned. But the rifle still failed to function smoothly and tended to jam whenever an effort was made to transfer a shell from cartridge clip to breech. The whole process of cleaning and oiling was repeated without much, if any, improvement.

Next morning, anxious to get out on the tundra, we were busily making a quick breakfast when Mary broke in, interrupting the proceedings. "Just a minute, Cy, I think I heard a white-crowned sparrow."

We flung open the door. A quick flutter, and there was her bird, perched on a willow twig a few feet from the step. The brilliant black and white stripes on the head, pink beak and flawless pearl-grey underparts of this large, handsome sparrow served to identify him positively. He was joined in a moment by three smaller sparrows, streaked brownish on their sides and backs, with rich chestnut caps and greyish white underparts, each sporting a central dark stickpin. They were tree sparrows, having migrated all the way from the central or southern states of the United States.

Mary darted inside, to return in a moment with a handful of rolled oats and cornmeal. She sprinkled this offering in a hollow in the side of a decaying snowdrift and in a trice had a dozen white-crowns and tree sparrows feeding almost at arm's length. Still more subjects to observe.

"There's a new one, Mary. He shouldn't be here."

"Which one?"

"The fellow with the black crown, face and bib. He ought to be a good deal farther south, beyond the tree line."

"It's a Harris's sparrow, isn't it?"

"Dead on! But he shouldn't be way up here."

The snow geese were still pouring through, a dark gyrfalcon occupied our flagpole, and from the snow-covered river below the loud calls of a large flock of tundra swans serenaded us. The ptarmigan were busier than ever and they were so numerous that we decided to carry out a study of them in order to learn something of their habits. Most of the morning was spent staking out an area in the willows a mile long and a quarter of a mile wide. Red ribbons, placed at intervals, would enable us to determine territories and pinpoint interactions observed.

Then, slinging the rifle over a shoulder and slipping cameras into our packs,

New arrivals to the delta were watched for eagerly.

we were off to see what else we could turn up. Within minutes, we were almost upon a pair of ptarmigan feeding upon willow twigs and new silvery pussy willows, just breaking out from their polished chocolate-brown buds. The male, in full breeding plumage, made a striking image in the viewer of my camera as the shutter was released. The female had begun to change her white winter livery to the grey, brown and silver plumage of midspring, with the first few darker feathers showing. Again, the male tended to be the more conspicuous, often frequenting small open areas while the hen never left the scant cover offered by the bare willows.

I began to puzzle about the significance of the difference in behaviour of the two members of the pair and the earlier change in colouration on the part of the male. Certainly it was important that the male assume his bright courtship colours early if he were to attract a mate in the midst of much competition, yet this change rendered him more vulnerable to predators while the tundra was still heavily blanketed with snow. These two elements appeared to compete with one another.

Suddenly, a male just ahead of us shot swiftly and noisily into the air, pursued a split second later by another male, and the chase was on. North for about a half-mile, then west across the river, over the breadth of Study Area Island to Fox Den Island and then back in a great curve to land with a loud clatter almost at the point from which they had departed some three miles earlier.

Now, they began running across the snow in a line parallel with one another, heads lowered. They paused at intervals, faced one another and growled, "go-back, go-back, go-back!" followed by "g-dout, g-dout, g-dout!" The invisible line between the two was undoubtedly the mutual border between their adjacent territories. The establishment of territories is very important in several members of the grouse family if uninterrupted mating is to take place within these limited preempted areas.

On the way back to the cabin, a very light male marsh hawk cut across in front of us on dihedral wings, its snow-white rump-patch clearly visible. Our diminutive, red-capped tree sparrows ornamented the tops of willow thickets, their wingbars very apparent, and our first gull, a very large nearly white one, flew over. This would be a glaucous gull, the largest member of its family in the world. There would be many of them here later, breeding on some of the islands, we knew. About halfway back, two pale short-eared owls were seen, flapping lazily over the willows with deep, slow wingbeats. One of them drove towards the other, nearly brushing its wings, and the soft, resonant, closely spaced courtship notes, "too-too-too-too-too," drifted back on the stirring north breeze.

It was becoming increasingly clear that many species of birds had evolved the pattern of breeding in more northern latitudes where the pressures of finding adequate food for nourishing their broods would be less than those on the wintering grounds farther south. While the supply of nutrients there might well be sufficient to sustain the adults, the rearing of new families of youngsters would pose greatly increased demands upon the resources of the environment.

We had barely started on again when Mary suddenly pulled my parka sleeve. "The bears!" she cried. "They're back!"

"Where?"

"Down on the river. Look!"

The two grizzlies were moving steadily along over the hard wind-packed snow on the river. Their dark forms loomed as large as elephants in the dead-white expanse of snow around them. I didn't like the direction they were taking.

"It looks as though they are heading straight for the cabin. And we'd better try to beat them there."

We tried to hurry forward, keeping the bears just in view over the rim of the river bank, but the thawing snow was deep and softer among the willows and we were clearly losing ground.

"I'll just pump a shell into the breech. We may be able to scare them."

But as so often before, the mechanism jammed and the shell would neither go in nor out. We waved our arms and bellowed and the bears took off at a gallop, vanishing quickly below the steep bank, but apparently still heading for the cabin. We moved as quickly as possible to the very edge of the bank from which we could get a better view of what was happening below. The bears were going at quite a clip but had changed direction and were heading downstream from the cabin site. We finally arrived at base camp completely out of breath.

"Thank heaven they didn't get here ahead of us," Mary breathed.

It was clear that we needed a gun that we could rely upon in case the bears gave us trouble at the cabin or we happened to surprise a sow with a young cub

The Arctic ground squirrel roused from hibernation beneath the deep snow.

in the tangled willows during our stay. The logical strategy was surely to get in touch with Jerry in Inuvik and have him send one out should a plane happen to be coming our way. We could get a message out by radio, of course, I thought, and accordingly switched on our set.

"Anderson River calling Inuvik. Anderson River calling Inuvik. Come in, please."

Hoots, squeals and grunts. Nothing more.

"Anderson River calling Inuvik. Anderson River calling Inuvik. Come in please."

Still nothing. Maybe the operator in Inuvik was out to coffee. I'd try later. This went on for eight days, several times a day, without results. On trying still once more after supper on the eighth day, a strong voice came through clearly.

"You callin' Nuvik? Over."

"Yes, I am. Over."

"I hear you lots of times. You get through yet? Over."

"No, no luck at all. Who are you and where are you located, please? Over."

"Reindeer herder, southa Tuktuk. I get Nuvik pretty good. Maybe I get through for you, eh? Over."

"That's very kind of you. I'd appreciate that very much. Over."

"You have message? Who for and what should I tell them? Over."

"I'd like to get a message to the Wildlife Service. Do you know them? Over."

"You bet. For a long time. You want to get Jerry? Over."

"That's right. Over."

"You have a message, then? Over."

"Yes, could you please ask him to send a rifle out to the Anderson Delta if a plane happens to be coming this way? Over."

"You bet. You needa gun, hey? Over."

"Yes, I do. Very much. Over."

"Why? What for? Over."

"There are a couple of grizzly bears hanging around the cabin and my wife and I don't want any trouble. We expect to be here all summer. They could be a problem, no? Over."

"You from Outside, eh? Over."

"Yes, Alberta. Edmonton, why? Over."

"I know you from Outside, all right. Very easy to tell. Over."

"How did you know? Over."

"Because you afraid of grizzly bear. You don't need to be afraid of them. Over."

"We don't need to worry about grizzly bears, you say? Over."

"Hell, no! They only eat natives. Like me. Over."

I thought that over for a few seconds and then fired back.

"Hey, just a minute! How does a grizzly tell whether a person is a native or a fellow from Outside? Over."

"Hell, they can tell after the first three bites! But I'll try to get your message through. Jerry, eh? Over."

"Yes, and thank you very much. My name is Cy and my wife is Mary. Please call again. Early in the morning is the best time for us. Over and out."

"My name is Omingmak. You know, muskox. Early morning the best time to sleep. And dream about very nice grizzly bears. You should have them in for tea sometime. Tea, she's ver' good stuff. Over and out, too."

"What a card!" Mary exclaimed. "I'd love to meet him."

"That makes two of us."

The grizzlies came around again the next day, but circled widely before making off in the direction of North Bluffs, a prominent hill to the north. An hour or so later, we spotted them both on a slope to the east of the cabin. Working slowly and methodically, I managed to get a shell into the breech of the rifle, in case we needed it.

Some sort of strategy seemed in order. The bears must be very hungry after arousing from their long winter's sleep, yet food must be hard to come by on the open tundra so early in the season, we felt. The deep scars on the ceiling bore mute evidence that the cabin had been raided by grizzlies before. We concluded that they were likely to return. While we could keep a lookout for them during our waking hours when we were working in the immediate vicinity, they could easily make an attempt on our stores while we were sleeping or busy farther afield. The spike-armoured shutters on the door and lower windows could be put in place, but that still left access by the two somewhat higher and unprotected windows. We decided to booby-trap the latter.

We rolled large empty oil drums under these windows and stood them on

On occasion, a few timber wolves visited the delta.

end. On top of them, we stood smaller iron drums which had been used for transporting white gasoline for the three-burner cookstove. And finally, we placed on the very top a miscellaneous collection of smaller tins and pots that we felt must surely create an unholy, hair-raising clatter should the bears attempt to gain entry by scaling these unstable pyramids. The noise would be sufficient to raise the half-dead.

It seemed that we had hardly dozed off when a great metallic "boom" reared us into sitting position.

"Grab the gun! They're back!"

I quickly picked up the gun standing against the wall by the head of the bunk, hastily slipped the safety catch and headed towards the window from which the alarm had come. I peered cautiously out.

Nothing.

"They must have gone around behind," Mary whispered, at my shoulder.

I pulled on my boots and trousers, gently eased the catch on the door and slipped out carefully, gun at the ready. Nothing to the west. Nothing in view from either end of the cabin. Nothing around behind. The sound must have spooked them and they were off for parts unknown. On checking the pyramids of containers on the way back into the cabin, they appeared undisturbed.

"Curious!"

The experience was repeated a half-dozen times before we finally got up for a breakfast of delicious bacon and eggs, toast and steaming coffee. "You know," Mary said, "I just can't believe that those bears kept spooking the whole night."

"It certainly doesn't seem reasonable. Hey, I've just thought of something. We disturbed all of that snow under the windows when we rigged our alarm system. The fresh tracks of our friends have got to be there. Let's have a look. We might even be able to tell how frightened they were by the rate they galloped off. The tracks in the snow will tell us a lot."

There weren't any tracks! Other than our own. We stood there, side by side. Unbelieving. Puzzling. Ptarmigan were towering and calling in the background and the sounds of geese and swans wafted up from the river. The sun was well above its midnight low and definitely beginning to assert a little authority. The air was balmier than it had been of late. The cool of early morning was already beginning to dissipate.

"I just can't figure . . . " But a sudden very loud "bang!" smote our eardrums. It had surely come from one of the drums in front of our very eyes. Then another "crack", lighter and higher pitched, issued seemingly from a smaller container at the top of the pyramid. Looking quickly upward, we were just in time to catch a slight quivering in one of the cans.

"Some bears, eh!"

"But I . . . "

"Of course! The drums are now expanding with the higher temperature. The air cools off when the sun is very low in the evening, so they were contracting when we were trying to sleep. Now, they are expanding."

Mary's peals of laughter broke the stillness. "You're the first bear hunter I've ever seen out gunning for rusty old oil drums in his pyjamas!"

Meanwhile, the snow geese and swans continued to pour in, along with many more gyrfalcons, four golden eagles, two peregrine falcons, several pomarine, parasitic and long-tailed jaegers. Then a small flock of white-fronted geese, their high-pitched calls unmistakable. Then our first small flock of black brant geese, their dark bodies winging swiftly towards an area just across the river.

The brant geese came in from the north rather than the south. We had often seen them on the Pacific coast of British Columbia and knew that at least part of their migratory route lay along there. They then, according to various sightings, fly on northward, around Alaska and then along the Arctic coast before flying into the deltas of some of the rivers from the north. We often heard them before they were seen, identifying them immediately by their hoarse, throaty "krrr-onk, krrr-onk" calls.

"This is all far beyond my wildest dreams," Mary smiled, resting on the end of a strange log projecting at an angle above the surrounding willows, her field glasses raised.

"Dreams or nightmares? How about the grizzlies?"

"We figured from the beginning that there is always the hazard of sudden illness, accident or miscalculation," she replied evenly. "And our pilot said that he knew of no incident in which barren-ground grizzlies had attacked people. Say, isn't that a pair of old squaw ducks down on the ice? What a glorious land!"

I brought the birds in with my glasses. No doubt about it. The long, dark, slender pair of central tail feathers of the colourful male rose above the icy snow like efficient tapering daggers. His less colourful mate rested on her belly beside him. We were to see many of these attractive ducks in the weeks to follow, and

Fortunately, the grizzly beat a hasty retreat.

their captivating brass instrument-like "ow-oww-let" calls would ring across the delta a thousand times a day, never failing to lift high spirits another notch or two.

Almost together, we spotted one brown object pursuing another brown one across the snow in front of us. They hugged the surface of the snow, twisting and turning, round and round in tight arcs. At length, the pursuer caught up with the pursued and over and over they tumbled in the snow, biting, scratching and chirring.

"Ground squirrels! Arctic ground squirrels!"

We focused our glasses on them. Grizzled brown on the back, bushy tail, rusty head and underparts. Largest of the true ground squirrels and the only species on the Arctic Barrens and coast. The combat was serious. Wisps of fur drifted across the snow from the combatants and the snow immediately about them became spattered with bright blood. As we watched, a third squirrel appeared and lay demurely and quietly in the snow observing the struggle. Rival males undoubtedly fighting for possession of the smaller, slimmer female. At length, they all disappeared beneath the high snowbanks running up from the river, the female quickly slipping down a burrow ahead of one of the males. We followed and saw where they had dug up through the deep layer of snow from their burrows below. They couldn't have been better insulated against the low temperatures and driving winds of an arctic winter with forty feet of snow above their burrow.

The ground squirrels were to become regular visitors at our doorstep for a week or so before vanishing below for a period of about six weeks. There the young would be born, only to come up to the surface of the tundra after the eyes had opened and they had become fully furred. Then there was only time to forage and fatten up in preparation for the long period of hibernation to come.

"Still another way of surviving the rigours of an arctic winter," suggested Mary. "They couldn't possibly freeze in their burrows under all that snow!"

The Key

When we went out the next morning, a warm caressing breeze was blowing from the southeast and many of the large flocks of swans on the river were swimming. This was not breakup. The swans were floating easily and gracefully on runoff channels which had developed overnight on either side of the midline of the river. An increased volume of water from farther south had created tremendous pressure upon the underside of the very thick ice above the main stream near the middle of the river.

This ice had been forced upward and had ruptured in many places, releasing water through the cracks. However, the ice along the margins of the river was unable to rise since it was still firmly frozen and locked to the bottom and banks of the river. The freed water flowed into these relatively lower surfaces of the river where it formed growing runoff channels, deep enough to allow us to travel by canoe.

On looking across these runoff channels towards Study Area Island, we noticed concentrations of snow geese, resting quietly on their bellies or walking about in the snow. A few small, snow-free patches, covered with what appeared to be tufts of dead yellow grass, had begun to show on the otherwise snow-covered tundra. Couples of the white geese were resting either on or near these areas. While flocks of snow geese were still flying over from the south, the local population did not seem to be increasing accordingly. Undoubtedly many of the birds were merely passing through on their way to more distant nesting areas. We decided that it was time to go over and check Study Area to see how things were progressing.

It took some time to dig the eighteen-foot freighter canoe out of the huge snowdrift which had buried it, take it down from its rack, put the shutters back on the windows in case of a raid by the bears, throw in gear and paddles, and set out. Mary took the bow, pulling with a stout rope while I pushed from behind, and we began skidding the canoe out to the near runoff channel. Here the water was already a good foot in depth and we were able to paddle for a few strokes. Then over the ice to the second channel for another brief paddle. Mary climbed over the bow with the towrope in hand as I clambered out, the stern end of the canoe still floating. Suddenly the ice below gave way and I found myself floundering in the icy water.

Fortunately, I still had a firm grip on the vertical stern and managed to pull myself back into the canoe. The low sun was shining brightly and there was no wind. We decided to have a look around before returning to home base, so I removed my clothes, wrung out as much water as possible and put them on again. The initial chill of the wet clothing took my breath away at first, but shortly I began to warm a bit. The sun was welcome, and fortunately, I had soaked up little water above the waist. Should a cold wind come up, we would head back.

As we walked towards the geese, some of them took to the air, but almost always in pairs, apparently most of the courtship and mating having taken place before arrival on the breeding ground. Several of the pairs had already begun nesting activity on the small bare areas, pulling the grass and weeds together to form shallow cups. While we found no eggs that day, two days later the birds had begun to lay their eggs. In the short period of ten days which followed, all of the snow geese had laid all of their eggs and begun incubating them.

Nesting and egg-laying all in ten days, in the entire group. It did not seem possible. But this was not the temperate region far to the south, where such activities might be spread over weeks or even months. Ring-necked pheasants, introduced to Australia from Europe, may spread their breeding period over as long a period as six months. But here in the Arctic, the season was short and there was little time to waste or dawdle. Late-hatched goslings would not likely be sufficiently well developed to migrate before being frozen in with the advance of winter.

But how did all of this hang together? Ducks and many other birds that nest further south often lose their first clutch of eggs to coyotes, foxes, crows, weasels, raccoons, agricultural activities and so on. They usually respond by laying a second clutch, and if that, too, is lost, still a third but smaller one may be laid. I had discussed this same matter with Jerry back in Inuvik while we were fogbound. He told me that Arctic geese do not have the ability to start up the process of egg production a second time in the same year. Only so many eggs are developing within the bird's ovary during any spring and once they have been laid, that's it for the year.

"But what happens in these brief laying seasons if the birds should encounter a late spring when the nesting area is buried deep in snow, and there are no bare spots on the tundra where they can pull grasses and moss together to build nests for their eggs?" I asked.

"They just don't lay them, that's all. If bare ground is not available, they resorb the eggs, taking all of the rich nutrients that have been incorporated into them back into their bodies," he explained. "The eggs don't even leave the ovary."

Jerry had left me with a puzzle that night. Later, after photographing several of the goose nests with their complete complements of eggs, I remembered my earlier conversation with him. I spent a long time turning over and over the importance of the evolution of this behaviour pattern in arctic-nesting geese. If conditions were not favourable for egg-laying at the outset of the spring season, why did the birds not wait until the situation improved and then lay their eggs?

They clearly were unable to wait. Five or six eggs were in various stages of development within the ovary, and in the short period of time available, the geese must either lay or resorb them.

Then how was resorbing the eggs an advantage over laying them in the snow where they could not be successfully incubated? The possible answer was slower

Lesser snow geese arriving from the south.

in coming. If the eggs were laid in the snow during cold weather, the constituents would be totally wasted. If they were resorbed, the goose could make good use of them, nutritionally. Didn't this mean that the goose would be in still better physical condition than usual to migrate in the fall? Wouldn't she have a better chance of surviving the great sorting mechanism of migration and returning to attempt nesting the following year? As farmers often say, "Sometimes this is a next-year country!"

We were beaching the canoe after returning from Study Area when Mary spotted a wheatear resting on the tip of a snowbank, its white rump, black and white tail pattern, and rusty underparts crystal clear in the sunlight. The bird took us completely by surprise and transported us immediately to the Ngorongoro Crater area in Tanzania, East Africa, where we had last seen one of them.

While the nearly bluebird-sized wheatear breeds across the whole of northern Eurasia, it has also extended its range into North America, both in the west via northern Alaska and in the east, via Greenland. Both of these North American populations winter in tropical Africa, the eastern group travelling by way of Greenland, Iceland, Europe, Morocco, Senegal and Sierra Leone, the western representatives by way of Asia, the Cameroons and Tanzania.

The sighting was a stirring one. The thought of a three-ounce bird carrying out these prodigious annual migrations challenges the imagination. In addition, how could the wheatears possibly find their way over such a long route, involving changes in direction at the appropriate times? If navigating by the sun, how did they compensate for its relative change in position, second by second? The same

Nest and eggs of the lesser snow goose.

problem faced them if navigating by the heavens at night with the pattern changing its relative location because of the earth's rotation. But undoubtedly there were other explanations, such as making use of the earth's magnetic field.

In the next few days, the runoff channels developed rapidly, providing us with opportunities to travel easily to many parts of the delta. The water on top of the ice was sufficiently deep to allow us to use the outboard motor, and we found this mode of travel much more pleasant than slogging through the heavy, moisture-laden snow on foot. Open spots on the tundra were expanding into sizeable areas of mud and grass, while small bare patches began to show at the bases of clumps of willows where the trapped snow had been deeper.

Shorebirds were making their appearance in numbers by now. Flocks of grey-breasted pectoral sandpipers, barred long-legged stilt sandpipers, buff-breasted sandpipers, robin-red knots, lesser yellowlegs, Hudsonian curlews and golden plover had made the long trip from South America. Black-bellied plover, semipalmated plover, ruddy turnstones, dunlins and semipalmated sandpipers came mostly from the southern, eastern and western coasts of southern North America. Thousands of northern phalarope and a light smattering of handsome red phalaropes had returned from the open ocean south of the equator, where they had wintered. In no time, the tundra vibrated with the twittering flight songs of scores of semipalmated sandpipers overhead, together with the soft, flute-like "poor me! poor me!" notes of the knot.

There were many unanswered questions. Why had these hordes flown so far north to breed? Did part of the answer lie in the fact that while the wintering

The snow geese began incubating with the laying of the final egg.

areas were able to support the large number of migrants during that season, they would be unable to support the same population, together with all of the pressures presented by the production of a new generation? Or had the birds been forced to migrate southward from these regions which had become increasingly hostile during the successive periods of glaciation? And, now that the glaciers had retreated, were they returning to their ancestral breeding grounds established prior to the glaciations? Or had the whole movement evolved in connection with the utilization of a rich, untapped food reserve available for the rearing of their broods? Was the ancestral home actually in the south, with these birds moving northward annually where conditions were more favourable for reproduction?

We wanted as many ptarmigan nests as we could find in order to get a pretty accurate picture of their nesting site preferences, the spacing of nests, the number of eggs laid and hatched, and their behaviour patterns which might shed some light upon their ability to survive in such an exacting land. The trouble was that the nests were very difficult to find since the birds were so secretive.

I remember having discussed this problem with a friend in the National Museum in Ottawa, who had spent many years in arctic research in connection with birds breeding there.

"Just how do you find ptarmigan nests in the Arctic?" I had asked him.

"I have a method that paid off for me, but I rather doubt that you'd like to adopt it as a regular strategy," he replied, smiling broadly.

"But you say that it works, didn't you?" I pursued.

Black brant geese nested nearer the mud flats.

"Not quite. It worked for me once. After hunting for nine years." His smile had broadened.

"Would you mind telling me about it, please?"

"Not at all. I was hiking across the tundra one day, on the lookout for nesting birds, when I spotted a very striking clump of woolly lousewort in full bloom. I took off my pack, laid it down and dug out my camera to take a picture. The brim of my old beaten hat cast a shadow on the flower when I was trying for a close-up so I tossed it aside on the tundra and took several shots of the flower. Then, having finished with the photography, I stowed my camera away, settled my pack on my shoulders again and stooped over to recover my hat. And when I picked it up, a ptarmigan exploded into the air from beneath it. My first ptarmigan nest! With ten handsome eggs! After nine years!"

"All that time to find your first nest?"

"Yes. Don't you think that's a pretty good method of finding them?" His eyes were twinkling.

"I'd like to multiply your magnificent score by twenty-five," I laughed. "And I'm not going to have two hundred and twenty-five years to do it! Nine times twenty-five is two hundred and twenty-five, isn't it?"

After many days and a great many more hours of concentrated field work, Mary and I had found only three ptarmigan nests by May 31, each with a single

Male willow ptarmigan in full breeding plumage.

egg. It was all very frustrating. The birds were nearly always seen in pairs, moving leisurely about in what we felt were their territories. We backtracked, round and round, in and out, following their imprints in the snow for miles in the hope that on occasion a nest would be found along the way. We had observed the fascinating courtship, culminating in mating in several instances. Courtships in which the hen flattened herself on the snow, at which the cock approached her obliquely with mincing steps, tail broadly fanned, neck bowed, scarlet combs erected on either side and wingtips dragging stiffly in the snow. Should the hen lie still, the cock treaded her briefly and then stepped down and described a circle or two about her as she fluffed her feathers and rose to resume feeding.

But should the prostrate hen swing her head and neck from side to side while uttering a soft mewing sound, mating did not occur. The decision to mate or not to mate seemed to be hers alone, which reminded us of the mating behaviour of blue grouse and Franklins farther south. There, while the males have a low level of discrimination and will attempt to mate with any of several species, the hens will respond only to the advances of the appropriate male. The responsibility for the preservation of the identity of the species rests largely with the behaviour of the female.

While there were many interesting stories in the ptarmigan tracks that we followed, there were few nests. The first nest was discovered by observing a pair continuously for about eight hours. The hen finally settled at the base of a thicket while the male perched watchfully on another willow about twelve feet away. The hen at last stood upright and began making pecking motions fro and aft, placing wisps of grass in the cavity in the snow. When we checked after she had

The female willow ptarmigan already in summer plumage.

sauntered off accompanied by the cock, we immediately found that the cavity contained one warm egg, lightly concealed by the dead stems. A red-letter day!

The second nest was discovered in a slightly different way. We had been following a pair for more than half a day when we both turned with our field glasses to watch the magnificent flight of a trio of parasitic jaegers, manoeuvring above us at breakneck speed. Apparently, two males were vying for a female. We had never before observed such speed, precision and timing in the flight of birds. The following birds seemed to anticipate every move that the bird up front was about to make, and the pattern of their manoeuvres as they turned, twisted, rose and fell was as tight as one could possibly imagine.

When we turned back to our ptarmigan, the male was alone; the female had disappeared. An hour's careful searching of the immediate vicinity turned her up, squatting in a deep cavity in the snow at the base of another clump of willows. When she left twenty minutes later, this cavity, too, contained a single, blotched egg. The third nest was discovered in much the same way.

We needed many more nests if we were going to be able to do an acceptable study. The key to finding ptarmigan nests much more easily came unexpectedly. One day we were observing a pair of the birds in the willows. The hen was feeding as though possessed, leaping up for pussy willows, snapping off small twigs and buds and swallowing them as though half starved. The male was perched on a tall willow, constantly turning his head as though searching the sky. Suddenly a parasitic jaeger appeared and the cock emitted a brief "kuk! kuk!" at which the female froze, motionless. When the jaeger disappeared in the distance, the male emitted a soft, drawn-out "caaaahhhh!" at which his mate resumed her rapid feeding. After eight minutes on the feeding area, the hen suddenly ceased her foraging and took to the air. The male followed closely and I followed their course carefully as they flew in an easterly direction. The female dropped to the ground and was instantly lost in the willows, while the male alighted on a tall

The eggs of the willow ptarmigan were well camouflaged.

willow at about the point where she had vanished. Here, he appeared to stand guard alertly.

I turned to Mary. "I'll bet a dollar that she has a nest there."

"You'll have to give me very high odds before I'll take that bet," she responded.

Could this really be the nesting area? We waited a few minutes before walking quickly towards the perching male. On our approach, he stretched his neck upward and gave a series of loud warning notes, descending in scale and reminding us of an outboard motor engine being started and dying out. We searched the immediate area bit by bit and soon spotted the female, crouching on her nest. When we gently flushed her, the nest contained five handsome eggs.

Thereafter, we had only to watch for flying or feeding pairs. If flying, they were either en route to a feeding area or on their way back to the nest. If the former, we had only to wait a matter of eight to ten minutes, after which they would fly back to the nesting site. Using this strategy, we were able to discover no fewer than thirty-two nests by the end of the season and, more important, the nests of all of the pairs of resident ptarmigan in the area which we had staked out.

As we had so often found, there is always a key in such situations. The pressing problem is to find it.

The Impossible, Possible?

We awakened on June 2 to a tundra heavily blanketed with new snow and a sharp drop in temperature. It snowed all day, obliterating virtually all of the landmarks on the tundra. The white-crowned sparrows and red-capped tree sparrows turned up on our doorstep for a handout and were given handfuls of crumbs and cornmeal. There was very little else that we could do outdoors with the almost impenetrable whiteout. We brought our notes up to date, sawed a bit of wood and replenished our water supply by heaping the gasoline drum back of the stove with fresh, clean snow. Our ptarmigan were laying additional eggs at the rate of about one in thirty hours, and we wondered how they would respond to the abruptly changed conditions. Internally, eggs would be ready for laying. What would the birds do?

The snowfall let up the following day, but our world had been transformed. The bare scars left on the tundra by days of thawing had all miraculously healed, having been mantled with a thick blanket of white glistening snow. A few exploratory steps into this new world revealed that rubber boots, open at the top, were useless for tramping around in the depths of snow which had accumulated. We needed our high waders, as well as the darkest sunglasses we had brought, since the brightness of the landscape under pearl-grey, misty skies was almost blinding to the naked eye. We set forth, barely able to make out the faint outlines of the closer islands on the delta. The tundra looked as though spring had foresworn her earlier promises and veiled her shame beneath a deep robe of ermine.

The soft snow was almost up to our hips as we trudged about the ptarmigan meadows looking for some of our nests. Each had been marked with a wisp of bright red surveyors' ribbon tied to the topmost twig near the nest, but only a very few of them were in view. The others were buried out of sight. The ptarmigan had already been abroad, their feathered feet leaving clear imprints where they had wandered about. The sets of footprints were usually in pairs, one bird travelling along with the other quite closely. On occasion, parallel furrows in the snow, five on each side of the line of tracks, told us that the cock had been dragging the tips of his primaries in courtship. At another point, the rounded impression of a ptarmigan's body flattened in the snow. Here, the two sets of tracks converged and the impression of the body lay at the centre of ten radiating shallow furrows, left in the snow by the tips of the male's primary wing feathers. The pair had mated.

We were heading, as best we could, in the direction of the spot where we had found a nest containing two eggs two days earlier. The going in the deep snow was heavy but we finally arrived in the general area where we thought we had found the nest. At this point we were stumped. The snow was well above our knees and only the very tips of a few willows were visible. It seemed utterly hopeless to look for a ground nest in the vast expanse of white, but the cabin was in view and we seemed to be in approximately the right place. I paused, breathing heavily, while Mary worked her way slowly forward on my right.

"Hey, there are fresh tracks over here! Looks like two birds, too," Mary called.

She was right. We followed the erratic, often circling, trail left by the birds, wishing that our feet were as effectively fitted with snowshoes as were those of the birds. Suddenly, we came upon what appeared to be a sizeable excavation in the snow cover.

"Do you think that this could be the site?" Mary queried excitedly.

"I'd guess that the chances are mighty slim, but here, hold my mittens while I dig."

I scooped out most of the snow with my bare hands, enlarging the cavity as the soft edges kept caving in. Finally, with my arm in to the shoulder, I felt bits of vegetation at my fingertips and then, eggs. Three of them. I drew them carefully from the snow, one at a time. Two of them had been inscribed "1" and "2" with a felt pen; the third was unmarked.

"Well, I'm absolutely dumbfounded. She found her nest in all this!"

"And I can hardly believe it, either," Mary chimed in. "Our nest! The one we were looking for!" We marked the third egg, carefully replaced and covered them all as we had found them, and departed.

How could the hen possibly have found her nest? All landmarks in the immediate area around the nest had disappeared, yet she had known where to dig. Could this be just an isolated incident, a stroke of luck? Two other similar records that day eliminated the sheer-luck hypothesis. Another female ptarmigan had attempted the same feat, but fell short a scant six inches of the nest, while a third was apparently unable to dig through the tough overhanging willow stems and left her egg deposited in a small crotch immediately above the nest. We were to find another four similar situations after the snow melted — situations in which ptarmigan hens had attempted to lay additional eggs during the storm, but had missed the nests by distances varying from inches to three feet.

So it was clear that our ptarmigan had a well-developed capacity to pinpoint their nests with remarkable precision, even in the absence of landmarks. While their scores were not perfect, they were very impressive.

Earlier experiences with other species seemed to indicate that many birds possess astonishing memories for spacial relationships. This is not to infer that this is the only mechanism used during long migrations. The latter is patently not true, since cowbirds, reared in the nests of other species, and the first-year chicks of such species as golden plover are able to make their way to the wintering grounds without having made the journey before and in the absence of experienced guides to conduct them.

Most students of bird behaviour shortly become aware of the ability of adults to forage widely for food and then return unerringly to the nesting site even though it is located in the depths of heavy forest, tangles of marsh reeds and rushes, or dense thickets of scrub. In such instances, they might well be making use of visible landmarks. On our farm, south of Edmonton, Alberta, a diminutive least flycatcher returned, probably from Central America, to nest on the identical aspen stub for three years in a row. The used nest had been removed after each breeding season. She built a new one on the same site. What role did memory play here?

While working with piping plovers many years ago, Mary and I had live-trapped our birds and banded them. One of the females occupied the identical scrape for two successive seasons. The latter involved a unique set of circumstances in that we had placed a wire cage over the incubating bird in order to protect her eggs against their being raided by marauding crows. The strategy worked concerning the control of predation by crows, but much more interesting was the fact that the shorebird returned to lay its eggs in the same scrape the following year. And this despite the fact that the cage was still in place. The nest in question was located on a very small island in the middle of a lake, and a violent storm had prevented us from removing the cage at the termination of the breeding season the year before. Without invoking remarkable memory, how else might one explain such events?

The weather improved rapidly after the storm, and the snow began wasting away before our eyes. A quick visit to Study Area Island revealed that some of the goose nests were still buried beneath a layer of snow. On uncovering the eggs, we noticed that a few of them were frozen into the nesting material. While some were split with the frost, others appeared intact. We noted the condition of such eggs, marked them briefly with a felt pen and recorded such data in our notebooks. We were later to note that some of the eggs which had been frozen, but not split to the extent that the shell membranes had been ruptured, hatched successfully. Those in which both shell and membranes had been ruptured failed to hatch. The eggs and chicks of at least a number of arctic nesting species are unbelievably well adapted to chilling, even in the advanced stages of incubation.

Both Arctic and red-throated loons flew over daily, while slim, buoyant Arctic terns, with their bright red beaks, long slender tapering wings and scissor-like forked tails, perched lightly on bits of bare driftwood or drifted swallow-like overhead. The Arctic terns were heartwarming visitors to our tundra after their long migration across the northern Atlantic and down the full lengths of Europe and Africa to regions inside the Antarctic Circle before swinging westward.

They then flew northward to Arctic and High Arctic regions by way of South and North America. After leaving the Arctic · at the time when the twenty-four-hour day had barely begun to shorten, they had made this seemingly impossible journey to the Antarctic where again, they found themselves in a region enjoying twenty-four hours of daylight. The terns before us had the distinction of being members of that species which experiences more hours of daylight than any other bird in the world.

Water levels were rising on the delta. The runoff channels were expanding in

Breakup occurred on May 25.

depth and breadth as the snow melted and as additional volumes of water were added farther south in the upper reaches of the river. Breakup occurred on the early morning of June 13 when a low intermittent grinding sound assailed our ears on awakening. Running to the window, we could see that the entire breadth of the river in front of us was on the move. We dressed quickly and rushed outside to witness the annual event. The water in the main channel was creeping inexorably up the bank, bearing with it great masses of shimmering ice. Pressure from behind often heaved enormous slabs into a vertical attitude, towering high into the air before crashing down on the moving surface below and shattering into a halo of glittering splinters mixed with foam. Still other blocks of ice were forced high upon the riverbank and left stranded there when the waters receded. In the next few days these abandoned blocks would candle vertically, forming incredible shimmering sheaves of slender crystals, conjuring up exquisite, exotic flowers.

The progress of the enormous ice field was not steady, but hesitant in that temporary jams developed and persisted until increased volumes of water built up and floated them seaward. At times, the ice cakes were interspersed with dark masses of twisted debris and fallen trees which had been gathered upstream by the river in spate. The huge blocks of ice that were pushed well up on the shoreline and left there would supply us with fresh water until the river subsided and cleared. The main channel in the centre had broken through completely by the end of the day and carried only sparsely scattered, freely floating fragments of ice.

Willows were bursting into full bloom, shedding miniature clouds of yellow pollen when brushed ever so lightly. Black brant geese were busily nesting on the mud flats just above high tideline in areas slightly lower in elevation than the sites chosen by the snow geese, which had arrived ten days earlier. About sixty tundra swans were foraging in the open river, dredging up pondweed that had been preserved in the bottom ice over winter and now made available to them with the disappearance of the ice.

The loud calls of the swans, coupled with those of loons, geese, terns, old squaw ducks and courting shorebirds produced a rich medley of sounds, eminently pleasant. Numbers of American widgeons, resplendent in their white crowns and green cheek-patches, mingled freely with the swans, swimming in close company with them. Was there a reason for this? On closer observation we noted that there was hardly a swan without its coterie of ducks. Then we noticed that no sooner did a swan's head rise above the surface, trailing a straggling ribbon of aquatic weed, than the widgeons were in for a share of the spoils.

We were reminded of the behaviour of silver-backed jackals in East Africa, snatching bits of the carcass of a freshly killed zebra upon which a pride of lions was feeding. At times, a duck pulled strands of a string of weed from a swan's bill before she could swallow it. Looking at these graceful swans with their wonderfully arched necks and bright plumage only served to reinforce the hope of being able to photograph them at close range.

However, I recalled a conversation that I had had with an acquaintance, George Simpson, regarding the possibility of photographing tundra swans at close range. George was a stocky grizzled outdoorsman who had spent a dozen seasons at various points along the arctic mainland. I had seen many of his still shots of barren ground caribou, Arctic fox, tundra wolves, ptarmigan and other species. They were superb.

I had mentioned that I was particularly interested in photographing tundra swans at the nesting site, especially at the time of the hatch, since there was still some controversy as to whether the cygnets left the nest followed by the adults or whether the parents actively led the cygnets away when the time came. Nor was there much known about the cob's role in the incubation duties and the early care of the young. But most of all, I wanted to attempt to capture details which could not be fully appreciated at a distance.

"You'd be wasting your time, that's all," he had advised.

"It won't be possible to photograph them at the nest, then?" I had pressed.

"Not one chance in a million. They're much too wary. They are gone almost as soon as you raise your field glasses. You'd need a camera with a 2,000-mm lens. Even then, the shimmering heat waves off the tundra would kill your shots. I'd advise you to try something simple. Like getting a super closeup of a sasquatch with his girlfriend!"

I respected George's opinion. He had had a great deal of arctic experience, but nevertheless, I was determined to try the tundra swans should an opportunity present itself.

Shortly after breakup one morning, Mary and I were hiking along the brow of South Bluffs, a steep, rounded hill overlooking the main arm of the river and many of the islands in the delta. Immediately across from us lay the south end of

Study Area Island, which extended northward to Gull Island, already populated by a goodly number of glaucous gulls that would later provide us with a reliable supply of fresh eggs. A narrow, shallow channel separated Study Area from Fox Den, a still larger island extending in a southwesterly direction. The diminutive Boat Island, very low and almost barren of vegetation, rose barely above the swollen river.

For a moment, the tinkling notes of Lapland longspurs above us broke our concentration, but then we turned again to the scene before us and began to sweep it carefully with our field glasses.

"What's that white spot on the other end of Study Area, just this side of that small pond?" asked Mary.

I swung my glasses in that direction and the white spot came into view. I brought it into sharp focus and studied it for several seconds. The shape was quite long and narrow and the colour, uniform white. It didn't move.

"It looks a little like a swan resting on its belly, but I can't be sure. There's another bit of white showing through that tangle of willows to the right. Let's go out to the edge of the bluff where we can get a better view," I suggested.

At the edge of the bluff, we must have been somewhat highlighted against the pale sky behind us.

"They are swans!" cried Mary. "Look, they're taking off!"

The two swans were running across the tundra with wings pumping hard. In another moment they were in the air, long tapering pinions appearing hard-pressed to overcome the force of gravity. But once the great birds were under way, the wingbeats became shallower and they flew swiftly and easily. They were heading directly away from us and we watched them out of sight over the ice pack to the north. The "white spots" had disclosed their identity.

"Do you think that they may have a nest over there?" queried Mary, a light frown creasing her forehead.

"I doubt it. It seems a bit early but that might not be so. I've never worked with tundra swans. They could have been resting after feeding on the weed in the river."

But when we spotted them again in the same place two days later, we decided to investigate. We crossed the river in the canoe, having planned to beach at the point farthest from the two birds so as not to disturb them unduly. We pulled the canoe well up on the bank in case the tide should rise in the delta and float it away in our absence. We slipped cautiously up the bank to peer over the rim, but had barely reached a point at which we could get a view of the general area to the north when the swans took wing. Only our heads could have broken the skyline, and we were almost a mile distant.

"Man, George was certainly right. I've never before tried to work with birds as wary as these. Prairie falcons and eagles are a cinch in comparison!"

"It looks as though he was right in advising us to forget the idea of photographing them at the nest," Mary offered.

"It looks like it, all right. But let's go and check anyway, now that we are here. They do seem to be on territory."

We had made our way almost to the north end of the island before we spotted what appeared to be a large mound of dried grass standing above the flat contour of the island. Hurrying forward, we were surprised to see that it contained five very large eggs, the colour of rich farm cream. They practically filled the large depression in the top of the nest. A few wisps of white down had been added to the lining of yellowish grass which had been plucked from a sizeable bit of the tundra immediately surrounding the nest.

"And what do we do now?" Mary ventured, shrugging her shoulders resignedly. "The swans are nowhere to be seen."

"Well, I'd like to put up our blind and see whether they will accept it or not."

"But surely not right here. They would abandon the nest for sure!"

"I wouldn't take that chance, either. But I'd like to try putting it up away back by the canoe, initially."

"But they flew off the minute we raised our heads above the bank," Mary reminded me.

"Yes, they did. But I'd like to try starting from back there. If they refuse to come back, we'll take down the blind and admit that Simpson is right. It's not possible to photograph tundra swans successfully at the nest."

We went back to the canoe and across the river to fetch the khaki-coloured blind. The slender corner posts were pushed into the soft mud and down to the permafrost and the canvas slipped over them. It blended pretty well with the dead vegetation around. The flaps were tied down securely so that they wouldn't ripple in the wind and a guy rope was added to each corner for extra support.

"How long are we going to give them?" Mary asked, anxiety creeping into her voice.

"Well, the temperature is clearly above the freezing point and Jerry said that goose eggs can take a good bit of chilling, especially in the early stages of incubation. That could apply to our swans here, as well. Let's try it until tomorrow morning and see what happens. If they can't bring themselves to accept the blind by then, we'll scoot across and take it down."

As soon as we rolled out of our sleeping bags the next morning, field glasses in hand, we flung open the cabin door and looked across the river at Study Area.

"One of the swans is back!" Mary sang out, grinning from ear to ear.

"They're both back! The other is standing half behind that clump of willows near the small pond. The cob's on the nest. Remember, he was quite a bit bigger and his head and neck showed more rust stain from the water."

"The male's on the nest all right, but I thought that cobs aren't supposed to incubate. Just the pen."

"That's what the books say. Hey, watch! There's another swan coming in. It's landing on the pond."

The third swan was backpedalling furiously with his great wings and then splashed down, leaving a silvery wake behind. The resident cob was already off the nest and rushing across the water to intercept the stranger. The two came together, necks outstretched and beaks open. They began pecking viciously at one another, calling wildly. In a moment the resident male knocked the interloper off balance with a powerful wingbeat and then proceeded to stab at its head with swift rapier strokes. The stranger broke free and beat a retreat across the water with flailing wings, the other close behind. They took to the air with the trailing bird so close that he plucked several feathers from the other's tail. They vanished over North Bluffs, still scolding. Meanwhile, the pen, unnoticed by us, had settled over the nest, incubating.

In a minute or two, we could see the resident male returning. As he landed near the nest, the pen rose, slipped down from the nest and both faced each other with heads raised and pinions extended at the sides. Their voices rose together as though calling in triumph. Then both strode off to the pond and began swimming synchronously, side by side, with necks smoothly curved and heads dipping to the water and rising again in perfect unison. It had been a striking display. At length, the cob conducted the pen back to the nest, where she settled and wriggled comfortably over the eggs. The cob strode off a short distance and then squatted on his belly, neck arched downward with bill resting on the lower breast.

The blind remained in place for the next two days. Would they come to accept it as a normal part of the environment, or was that asking too much of these wary birds? On the third morning, the swans were still in evidence and behaving as usual. Having decided to chance it, we crossed the river and began pulling the corner poles of the blind from the sticky mud. Both swans remained near the nest until we started moving forward, the blind held between us. They then took to the air and began circling the general area as we strode some sixty steps closer to the nest, pushed the posts into the mud again, squared away the blind and left. While climbing the bank on the cabin side of the river, we watched the swans as they began to lose altitude. They landed on the small pond near the nest and waded to shore. The cob conducted the pen to the nest and then remained on guard nearby.

Next morning, the cob was incubating while the pen dabbled in the pond. Since they did not appear unduly alarmed by our interruptions, we decided to move the blind still closer. This time, the pen flew off as we reached the blind, but the cob failed to leave the nest as we moved in another sixty strides or so and again secured the blind in place. He did not flush until we had moved a dozen steps still closer on our way to the nest. It was imperative that the eggs be checked regularly if we were not to miss the hatch. The cob had flown to the narrow channel separating the two islands, where he joined the pen. We had not noticed that she had landed so nearby .

This strategy was continued until the blind rested barely thirty feet from the nest. It had been twenty-four days since we had first noticed the swans in their nesting area. On this, the twenty-fourth day, the birds had not bothered to flush at all. The incubating cob had waited until we were perhaps fifty or sixty feet away before he stood up, stepped deliberately down from the nest and walked slowly to join his mate in a shallow depression behind the nest. Both rested quietly as the eggs were checked. One of the five eggs had mysteriously disappeared, perhaps taken by a passing wolf or fox when the adults were briefly absent, feeding.

"I'd like to try some photographs today," I suggested. "The birds are certainly not alarmed at our presence now, and you never know how long our luck will hold."

"So what do we do next?"

"I'll hurry back to the canoe and get my camera and tripod, if you'll wait here. It will take only a few minutes and by that time the birds may be anxious to get back to the nest."

I returned and we both crawled into the blind. The camera was soon mounted on the tripod and focused on the nest. Mary then backed out of the blind and left, conspicuously waving her arms so that the swans would be sure to notice that someone had left the area. Mary was to take up position near the canoe. I glued my eyes to the tiny peephole near the protruding lens, wondering if the swans had noticed that while two people had entered the blind, only one had left. I was not left long in doubt.

Within minutes, the cob raised his head, peered in the direction of the blind and began to stride towards it. My heart raced as I watched him place one enormous black webbed foot ahead of the other, rocking a little from side to side as he progressed. The pen followed, a few paces to the rear. The cob paused momentarily at the edge of the nest, waiting for his mate to catch up. She then stepped carefully up onto the nest, arranging her feet carefully before settling. Then, with a shuffle and wriggle, she settled comfortably. From such close range, her head and long graceful neck appeared immaculate beside that of the iron-stained cob. The yellow spot at the base of the bill was as clearly visible as the reflected light in her dark eye. As I composed my picture in the viewer, the tall yellow stems of grass behind seemed to sweep up to the cold, white ice pack, lying just offshore. The cob turned slowly and strode off a few paces before settling on his belly with neck smoothly curved and eyes closed.

I remained in the blind for a full half-hour, giving the bird ample time to warm the eggs thoroughly. Then, about to shove an orange ribbon out through a hole in the back of the blind as a signal for Mary to return, I noticed the cob suddenly stretch his neck vertically and begin to stir. Then he rose to his feet and began walking towards his mate on the nest. On reaching the nest, he straightened his neck horizontally and thrust his head under her breast. She immediately stood up and stepped down from the nest. The cob stepped up and settled, taking her place. I could hardly believe my eyes. Both swans in the frame at once in an exchange of incubation duties! Speaking of golden nuggets!

Not wishing to interrupt the pattern of the cob and pen at the nest, I left him incubating for some time before signalling Mary. She returned quickly and entered the blind. The cob had remained on the nest until she was but a few yards away, when he stepped down and sauntered to a point near the pond where he waited, standing. We packed up the photographic gear, crawled out, tied down the blind securely and left as usual. The cob was back on the eggs by the time we had reached the canoe.

"How did it go?" asked Mary, unable to restrain her excitement.

The cygnets of the tundra swan had hatched and dried off.

"Absolutely marvellous! I even got the exchange at the nest. A red-letter day, if there ever was one."

"Then George Simpson wasn't completely correct?"

"We can't say that yet. Remember, he maintained that we could not photograph the adults at the nest at the time of the hatch. The eggs haven't hatched yet."

We continued checking the eggs each morning, as we felt that the hatch couldn't be far off. Two days later, there was still no sign of starring or pipping and we feared for their fertility. Late the next day, we were returning by canoe from working on one of the islands in the inner delta when, directly in front of us, we noticed a lone adult swan driving across the water on flailing wings. We cut the motor and almost immediately spotted three newly hatched cygnets in the water, only a few yards away. The little greyish-white fellows quickly ducked their heads beneath the surface and paddled like mad for several seconds. But when they raised their heads again, they found themselves in the same spot. The manoeuvre was repeated, but the cygnets made no headway at all.

"What funny little guys," Mary laughed. "They remind me of three plump ladies at the fitness spa furiously pedalling their exercise bikes and then checking the scales."

We started up the motor and pulled quickly away, giving the swan a chance to rejoin her cygnets. Then, almost in the same breath, we both exclaimed, "What about our pair of swans on Study Area?"

I revved up the motor and cut across the bay and along the bank of the main river to our usual landing place. The pen was sitting as we hurried forward, but she seemed more reluctant to leave the nest than before, and allowed us to approach within a very few yards before she stepped down and moved away a short distance. Three of the eggs showed very minute cracks at their larger ends.

"The eggs are not infertile. They are faintly but definitely starred. They are going to hatch!"

"How long do you think they will take?" asked Mary.

*The pen defended
her cygnets
vigorously.*

"I'm not sure. But if they are anything like those of geese, they could hatch sometime tomorrow. I'm not going to take any chances, so we'll come over first thing in the morning."

The first thing we did on the following morning was to focus our glasses on the north end of Study Area. Both swans were at the nest, but neither was sitting.

"Grab your camera and let's go! Breakfast can wait."

Both birds allowed us to approach without flushing, but they had moved off a short distance by the time we reached the nest.

"They've hatched! All four of them!" shouted Mary. "Did you ever see such gorgeous little chicks?"

Mary dropped to her knees beside the nest to get a closer look. The newly hatched chicks had almost dried off, but already the long silky down covering their bodies was beginning to take on the lustre and sheen of finely spun silver. The bills were ebony; the small webbed feet, a delicate pink; the eyes, dark and lustrous. I photographed them briefly, but did not wish to leave them exposed to the brisk air at this early stage for fear of chilling. So Mary quickly put me into the blind and departed. She would observe from farther back.

The adults were back within minutes, the female immediately covering her brood while the male remained standing a yard or two away. I stayed in the blind for the better part of two hours, observing and photographing them. The pen stood up in the nest at intervals and I could see the youngsters squirming beneath her breast. She finally rose, stepped carefully down from the nest, moved away a few yards and then stood looking over her shoulder in the direction of the nest. The cygnets squirmed and peeped for a few minutes before one of them climbed out of the nest cup and slid down the outside to the tundra. He landed on his tummy, but gathered himself up and waddled uncertainly towards his mother. A second cygnet followed suit. The final two, probably a bit younger and obviously not yet quite so strong, were even less sure of themselves. They stumbled, rested, stumbled forward again, but finally reached the pen. With the four around her feet, she waited patiently for several minutes before leading them very slowly in the direction of the near channel. The cob closed in from behind.

When I left the blind for a last shot or two, both adults turned defiantly towards me, raising their wings and hissing loudly. They were prepared, it seemed, to defend their brood against all comers. I drew back and watched them make their way to the water's edge. There they launched and moved away slowly with the current. The cob was in front now, with the pen bringing up the rear. The cygnets were in between.

"They are on their way," Mary breathed at my shoulder.

"They are that," I agreed. "I didn't see you come back."

"Did you really want me to miss the last line of the last chapter?"

Past and Present

Once most of the snow had disappeared, the solution of the riddle concerning the strange ends of logs protruding at irregular intervals above the tundra became apparent. It became clear that much of our ptarmigan study area above the river on the east side of the delta had been used in the past by Eskimo for a special purpose. The area had been an Inuit burial ground. While making our daily rounds to check nests south and east of base camp, several of these Eskimo grave sites were nearly always in view.

Since this is the land of the permafrost, it had not been possible for the Eskimo to dig graves in which to lay out their dead. Instead, a custom had evolved which was much more practical in the light of prevailing conditions. The bodies of the deceased were simply laid out on high ground sufficiently far back from the river to avoid flooding and erosion of the sites. Then large driftwood logs were either carried or dragged by dog teams from where they had lodged at the time of flood and piled neatly over the corpses.

The prized possessions of the departed were then laid on the logs. These included such items as the wooden runners of their sleds or komatiks, the running surfaces ingeniously faced with whalebone for longer wear. The lengths of whalebone used for this purpose had first been wrought into the appropriate shape and size by hand. Then small holes had been somehow drilled through the bone at desired intervals and whalebone pins fashioned to pass through these apertures and into the solid wood of the runner. Thus the two surfaces were securely bound together. The carefully worked wooden keel of a boat, together with paddles and simple rudder, had been added to one of the graves.

One of the most interesting items on another grave consisted of a nearly square piece of wood about eighteen inches on a side and two inches thick. The corners had been slightly rounded and a hole about three inches square had been carved from the centre, with pairs of small perforations near the corners. This was the shield behind which the Eskimo advanced when crawling over the sea ice to hunt seals or other polar animals. The side of the shield facing the seal was probably plastered with snow while the hunter kept his quarry in view through the hole in the centre.

Over the years, many of the logs covering the graves had slowly weathered and disintegrated, revealing portions of the bleached skeletons of those who had been buried. Examining some of the exposed long bones and skulls never failed to remind me of Shakespeare's Hamlet, holding and contemplating the skull of Yorick in the impressive gravedigger's scene.

Eskimo grave with the prized possessions of the deceased.

> "Alas, poor Yorick! I knew him, Horatio: a fellow
> of infinite jest, of most excellent fancy: he
> hath borne me on his back a thousand times."

Who were these people when they were alive? What parts had they played on the wide tundra's stage? How often had they struck out on uncharted courses in the dead of winter over unbelievably rough sea ice to visit friends many hundreds of miles away on High Arctic islands? Struck out without chart, map or compass? How had they come to this end? By lingering illness and incapacity? By sudden accident? Or, knowing that their time had come, had they remained steadfastly behind and welcomed the Dark Angel into their presence, serenely and quietly? What had been their final thoughts as gentle breathing slowed and the faint mist of the last breath rose to mingle with the dancing northern lights, low in the heavens above?

The old graves were not entirely devoid of life. Willows and Arctic rhododendron grew taller, here in the enriched soil. Male willow ptarmigan used the higher logs regularly as lookout posts while an ermine, already in summer fur, found the piles of driftwood a welcome retreat. He often greeted us by appearing suddenly from below and eyeing us curiously over a weather-beaten log. His animated face with dark sparkling eyes, long quivering whiskers, twitching nose and silver-margined ears disappeared and reappeared with disconcerting rapidity.

The sinuous body was pale chocolate-brown above with underparts whitish, tinged with pale yellow. The medium-length brown tail bore the typical black terminal brush. Considering the large size, this individual was clearly a male. But in addition, this fellow had the extra size so often found in northern races of birds and mammals which results in their losing body heat to the environment more slowly than is the case with their southern counterparts.

While observing this member of the weasel family there on the arctic tundra, earlier experiences with the same species much farther south came flooding back. In my mind's eye, another spring replaced this one and another ermine, the one before me. The earlier one had been live-trapped in early autumn and housed in a secure cage throughout the winter. It was a rather small female which as spring progressed became noticeably plump. This did not strike me as unusual and I had attributed her somewhat sudden corpulence to the abundant supply of mice and voles which had become so readily available as the winter snow disappeared.

I had kept her cage well stocked with these food items, which she eagerly snatched from my hand and stashed away in her nesting box. One morning about the middle of May, I had noticed her running about in the wire cage and seized the opportunity to open her winter shelter with a view to cleaning it out. To my great astonishment, the box contained a litter of very young, nearly naked weasels with their eyes still closed. Nine of them! How could such a small mammal possibly have such a very long gestation period? On telling fellow naturalists and zoologists about the experience, the responses were interesting.

"You mean you didn't notice the visits by her boyfriend?"

"Of course she escaped from her cage and it was awhile before you caught her again. Do you remember just when she was away?"

"I've been wondering when the idea of part-time live-in boyfriends would catch up with the rest of the animal kingdom."

"It did happen once before, you know. An immaculate conception."

"Not likely a test tube deal, I'd say. Nine's a bit much."

The weasel in question had definitely not been out of her cage since the previous fall. And if she could not escape, it was a dead certainty that a larger male could not have had access to her through the stout wire mesh. Careful examination of the entire cage failed to reveal any evidence of weakness or damage in its structure. But shortly thereafter, a zoologist from Montana, USA, unravelled the tangled threads and clarified the details of delayed implantation in the species.

Ermine, or short-tailed weasels, mate in the summer or early autumn, fertilizing the eggs in the female's reproductive tract. The early stages of embryo development begin but cease shortly, each embryo rounding up into a sort of cyst which remains within the female without becoming implanted in the wall of the uterus, thus arresting further development at that time. After the fall and long winter, the encysted embryos implant in the uterine wall and develop rapidly. Births occur in middle or late spring.

It struck me that this truly marvellous adaptation was particularly important to such forms living in Arctic or High Arctic regions. Mates are more readily available in the late summer and early fall when populations are highest, but the female is not faced with the enormous task of hunting large quantities of food during the arctic winter when food supplies are at a minimum.

Her young develop within her during the season when food, in the form of tundra voles, lemmings and the hordes of arctic nesting birds, is more readily available. After birth, the young still have sufficient time for growth and development before winter sets in again. As in many arctic forms, the balance is critical. Should the young begin development too soon, food supplies may be inadequate; too late, and the young may not be sufficiently well developed to survive the rigours of the ensuing winter.

Over the years, we had turned up many interesting nesting sites, ranging from the pair of robins that nested in an old delivery truck and took regular trips on a milk route to a pair of swallows incubating their eggs on a busy river ferry, a goldeneye duck that nested in a homesteader's chimney, a house wren in the pocket of a farmer's overalls. Then there were the two pairs of barn swallows that built their mud nests on either end of the shelf over a trapper's rude fireplace. The trapper was forced to keep the cabin door ajar throughout a dreadful mosquito season so that the swallows could incubate their eggs and rear their chicks. But these records pale before the nesting site that we found on the tundra.

One day, Mary and I were observing one of our ptarmigan hens busily foraging near an Eskimo grave. We were carefully timing its feeding period.

"I think I saw something move near the bottom of that old log that rises at an angle from the grave," Mary said. "Can you make out anything there?"

I quickly focused my glasses on the log just in time to see a small brownish bird hop upon the log from below. The white areas in the tail and dark streaking on the upper breast were quite apparent.

"Lapland longspur. Female, eh?"

"Yes, I get it now. I didn't get a good look at it before. The willows were in the way." As we watched, the longspur turned and ran quickly down the log and into the willows.

"You watch the ptarmigan and time her, Mary, please. I'd like to have another look at the longspur. I thought that the chestnut on the back of the neck was pretty pale, even for a female."

"Right, I'll stay with the female ptarmigan, Pinky, until she flies back to the nest." I continued watching for several minutes but the longspur did not reappear.

"Pinky has gone back to her nest. And the male's with her," Mary sang out. "Nine minutes, forty-three seconds this time."

"Good. I haven't seen hide nor feather of that longspur again, so let's see if we can flush her. She probably slipped away under the willows."

We walked slowly up to the grave, our eyes on the last spot where we had seen her. An exposed part of a skeleton, one paddle and a single sled runner were in view. Overhead, we caught the beautiful tinkling flight song of a male longspur. Looking up, we watched him descend almost vertically and land in the top of a willow clump a few yards away. From the river below, the distinctive, rich, harmonious "ow-owlet . . . ow-owlet" notes of the old squaw drakes rang out clearly. Mary had just stepped forward to seat herself upon one of the grave logs when a longspur burst into the air.

"She must have a nest right here," Mary said excitedly. "I saw where she got up."

"Where?"

"Almost right beside the bones." We both got down to check the small area more closely. Although several willow stems were growing a short distance away, there was little cover near the skull itself. She must have been scratching around for seeds or something, I thought, gazing reflectively at the vacant eye sockets in the bleached brain case, facing slightly to one side.

"Look!" said Mary. "There are some bits of dry grass sticking out of the back of the skull."

She was right. Several slender stems of yellow grass trailed from the round opening that normally admits the spinal cord. I carefully pushed my long central finger into the hole, exploring the hollow. It first encountered what seemed to be a ridge of matted grass and beyond that, a cavity with a layer of softer material. Reaching as far down as possible, my fingertip encountered a small, smooth, rounded surface. An egg!

"Our first Lapland longspur's nest!"

This was evidence enough that the nesting period of such small birds had arrived. We scoured the area and soon turned up more. Redpolls built their nests in the forks of willows and lined them with the white feathers of snow geese and ptarmigan. Tree sparrows and other Lapland longspurs constructed snug nests on the ground and also made good use of goose feathers for lining and insulation. The white-crowns nested similarly but used quantities of fine rootlets and grass between the cold ground and egg cup.

Meanwhile, glaucous gulls were setting up housekeeping on several of the islands. We chose a small island with about twenty-five nests of these numerous gulls and began gathering eggs from the nests every other day, since the eggs that we had brought along had all been used for meals and baking. These very large gulls normally lay three blotched and scrolled eggs, each nearly three times the volume of an average domestic hen's egg. We wanted them fresh, preferring not to use those with either partially or well-developed chicks in various stages.

"Eggs are eggs," Mary urged. "I'd rather have mine sans little chicks, fascinating though they are."

We soon had a good supply of fresh eggs. For the most part, they were perfectly edible and particularly good as an ingredient in the corn bread which Mary baked in our drum oven. However, once in awhile an egg or two would be a bit on the strong side, and we concluded that the bird had probably been feeding upon a dead fish or overripe carcass of some kind. While the normal complement of eggs in this species is three, as in most other gulls, some of our birds had laid more than a score when we later gave up the harvesting and permitted them to proceed with reproduction.

The manna from our gull island posed a problem when it came to storage. After an enormous amount of very slow work, we managed to dig a hole in the permafrost that would accommodate an oil drum. Digging was difficult because

A pair of golden eagles nested on the cliff, south of the cabin.

the consistency of the permafrost was about that of concrete. However, the saving grace lay in the fact that the digging surface thawed bit by bit as the hole deepened and we were able to gradually make headway. Finally, with the oil drum inserted into the hole in the tundra and a metal lid added from another drum, along with a bit of insulating canvas over the lid, we had a refrigerator.

When we were out on the southern end of the ptarmigan willows one day a golden eagle flew over, flying in the direction of South Bluffs. The bird dropped lower and lower and appeared to come to ground, with wings pumping heavily, just behind the near shoulder of the high ground. We decided to investigate. On making our way up the slope, we noted that fewer of the hitherto conspicuous male ptarmigan were perched high on the tops of willows. And then, here and there, we began to encounter masses of white feathers clinging to the tangles of dwarf willow. While the males that we did see were well into their brown summer plumage, the snagged feathers were white and often accompanied by the remains of ptarmigan wings. It was clear that the ptarmigan in question had fallen prey to some predator earlier, perhaps the eagle that we had just seen or the peregrine falcons observed in this general area on several occasions. On the whole, we suspected the eagle, since previous experiences led us to believe that peregrines prefer shorebirds such as pectoral sandpipers.

Earlier, we had almost invariably seen the male ptarmigan perched conspicuously on points of vantage, either guarding territory or on the alert for enemies as their mates fed between incubation periods. On the other hand, the hens tended to be much less conspicuous, both in colour and behaviour. Considering all of this, it appeared that the males were actually "sitting ducks" for such predators as eagles and peregrine falcons, while the hens were much less likely to be taken. If this were indeed so, what might it mean?

Did this mean that, having served their purpose in fertilizing the eggs, the males were now somewhat expendable? Did it mean, also, that by presenting themselves so conspicuously, they were taking pressure off the hens by attracting predators to themselves, rather than to their mates? After all, was not the future generation already assured by the fertile eggs residing within the body of the female or in her nest? Nor is the male necessary for the fertilization of successive eggs, since a single mating is sufficient for the fertilization of an entire clutch. It was an interesting speculation.

At any rate, the eagle did eventually flush in front of us and we discovered the nest. It was a large affair of coarse twigs and grass stems, located on a flatter area well up on the side of the cliff. Besides three creamy eggs, slightly blotched and spotted, it contained some remains of ptarmigan and old squaw ducks. Farther on, we flushed a peregrine falcon from an eyrie on a bare yellowish ledge of clay and sand. She left four pale, slightly rusty and faintly blotched eggs on the ledge. We were elated and hoped that we might be successful in photographing them at the nest following the hatch.

On the way back to the cabin, we stood watching a hen ptarmigan feeding while her mate perched watchfully on a grave log close by. Then, unexpectedly, we found ourselves in the midst of an invasion, an invasion of the area by no fewer than fifty male ptarmigan, all in a single flock. They tended to slink along the ground in a furtive manner as though unsure of themselves. Resident males rose to challenge them, driving them on and out of their territories. The interlopers stood their ground only very briefly. They certainly lacked the confidence displayed by resident males when challenging one another on borders of adjacent territories. We witnessed the same sort of invasion on two more occasions later, but were at a loss to explain the apparent surplus of unmated males.

Did it mean that, overall, more male chicks were hatched than females, that more female chicks tended to survive after hatching, or that winter kill was greater in females than in males? Other things being equal and considering the greater vulnerability of conspicuous breeding cocks, it seemed that there ought to be more females than males. Or was this merely a mechanism that had evolved in order to balance the loss of males by predation?

The strangers having departed, we began checking ptarmigan nests nearer the cabin in order to see whether any of the eggs were showing signs of hatching. By now, the hens were sitting very tight and no new eggs had been added for two weeks. At Nest 5, the hen was broadly flattened on the nest with neck low and head resting on the outer ring of breast feathers. She allowed us to drop on our knees beside her without flushing. The only movement was a single blink of the dark eye, reflecting a small spot of light. She created an altogether unforgettable image, a rich study in gold, silver and burnt umber.

The dark feathers of the crown and cheek were spotted with burnished gold while bright ripples crossed the sides of the neck, breast and back in a symphony of line and colour. The terminal margins of the feathers were gentle waves of silver. Despite the richness of colour, the effect was one of consummate camouflage, the sitting bird blending perfectly with its surroundings. I slipped my fingers gently beneath her breast, lifted her off the eggs and lowered her carefully onto the tundra beside the nest so that we could check the eggs for pipping. The eggs showed no signs of hatching and in a moment, the hen uttered a single low cluck and slipped back onto her eggs.

Two other nests were in like condition. Neither hen flushed into the air but, rather, ran off a few paces before settling on the ground while we did our checking. When we stood back a few steps, they returned to their nests and settled. One of the males was spotted resting under an overhanging willow. This was a change in behaviour.

On a bright sunny morning two days later, we decided to take a trip to Moose Track Lake. On the map, it appeared an intriguing-looking body of water in the shape of a moose's footprint. The trip would involve a lengthy leg up the main channel of the river before negotiating a narrow water course south and west of Fox Den and then crossing quite a large bay, marked on the map as a "shallow, narrow channel." We would have to be careful of grounding the motor in the mud and stalling it.

A gentle breeze was blowing out of the southeast and only a hint of pale cirrus lay on the northern horizon. Mary packed a light lunch and a thermos of coffee while I loaded two blinds into the canoe and we were off for the day. Taking it slowly, we managed the channels successfully and finally arrived at Moose Track. The lake itself was bordered with high, irregular banks which we felt must represent the polygons that we had seen from the air on our journey in from Inuvik. The surrounding area was lush and green, much of it so low and wet that we were glad of our rubber boots. With the canoe pulled well up on one of the banks, we were ready to strike out to see what we could find.

Almost immediately, we heard the rolling call of a whimbrel or Hudsonian curlew overhead and Mary spotted several glaucous gulls, apparently loafing on the banks along the lakeshore. We hiked to a small shimmering lake to the south, where we spotted a red-throated loon, floating head high, on the calm water. Was the loon alone or did it have a mate hidden away somewhere? We had hardly begun skirting the lakeshore when a subdued splash issued from the tall grass immediately ahead and then a series of ripples radiated out from shore.

"We've found its nest," said Mary excitedly.

The loon's nest was on a low mound just up from the water's edge. It had been constructed of wet, decaying vegetation and contained two large, chocolate-brown eggs with darker markings. The bird from the nest had joined her mate near the middle of the very small lake and both floated with heads up, apparently surveying the intruders. Their smooth grey heads, chestnut throats and slightly upturned lower mandibles were distinct.

"I'd like to give them a try later, but right now I'd prefer to scout around a bit and see what else might be here."

There appeared to be a somewhat larger lake beyond the one with the red-throats. We headed in that direction but had gone only about a hundred yards when a goose flushed from a stretch of low-growing but dense willow directly in front of us. It climbed wildly and then began a wide circle about us.

"Whitefront!" I shouted. "How lucky can we be? They're supposed to be here all right, but very widely scattered. Jerry was years stumbling upon one, remember. Just look at those black patches on its belly! No doubt about this one."

We hurried forward and immediately found the nest. Leafy, overhanging willows grew closely around it, effectively hiding the nest from most potential predators. Four large, creamy-white eggs lay in the centre, their less-rounded ends together.

"Are you going to try it?" Mary asked excitedly. "This may be our only chance."

"You bet I am! The loon will keep. Let's go back to the canoe and fetch the blind. I've never talked with anyone who has worked whitefronts at the nest, so we may not get a thing. But I still think that we ought to try. They may be very wary, you know."

When we returned with the blind, the goose was nowhere in sight. We had not seen the mate. We decided to begin with the blind well back from the nest and set it up at half-height, hoping that the goose might be more willing to accept a smaller foreign object in its territory. I got inside, seated myself on our old butter box and prepared for a long wait. I hadn't bothered setting up the camera, as images taken from that distance would be much too small.

I had hardly settled when there came a great rush of flailing wings overhead.I pressed my eye to the peephole and could hardly believe my eyes on seeing that both geese had landed near the nest. One of them walked immediately to the eggs and settled over them while the other lay down on its belly a yard or two away. They had apparently taken no notice of the distant blind. I waited half an hour for the goose to warm her eggs and then signalled to Mary. She hurried over, the geese flushing as she went by.

"Yes?"

"I'd like to move in at least halfway. The geese may be all right." And we moved the blind a good deal closer. Mary departed as usual and I felt near enough then for record shots, if nothing else.

I had no sooner attached the long lens and mounted the camera on the tripod when the birds returned as before. After taking several shots and leaving the birds at their duties for a short period, I again signalled Mary. We continued thus for perhaps two hours, at the end of which the blind was no more than a dozen feet from the nest. From this distance all the details of the bird's plumage were clear as crystal: the smooth, buff margined feathers running down the back and upper wings, the light barring on the sides of the breast, the pronounced blotches on the belly and even the very narrow line of white feathers at the base of the bill, responsible for its common name, whitefront. An experience of a lifetime and a great deal more than doubly exciting because the events were so entirely unexpected.

Since the photography of the whitefront had consumed so little time, we packed up the blind, returned to the canoe, had a welcome lunch and set forth again. The sun was warm on our backs and the light breeze, still out of the southeast, was caressing.

"What a day already," Mary offered, warm enthusiasm colouring her voice.

"Yes, and that's probably the most intriguing part of working outdoors like

Occasional northern harriers bred on the delta.

this. You never know what will happen next. The unexpected seems to turn up pretty often."

We were headed in the direction of the second lake, which we had seen earlier in the day. When we arrived at its margin, we could see that it was probably two or three times the size of the small lake on which the red-throated loons were nesting. We scanned the gleaming surface of the lake with our field glasses and almost immediately spotted another loon. Smooth pearl-grey head and hind-neck with the sheen of brushed velvet, deep ebony throat, brightly checkered back, dark pointed beak.

"Arctic loon, for sure. And all by itself. I wonder if it has a mate anywhere near."

"Well, let's look," said Mary, starting along the shore.

"After those whitefronts, I feel that anything could happen today."

We continued on, but stopped often to take a further look at the magnificent loon, floating serenely off the opposite shoreline. About an hour later, we were approaching our original point of departure when the head of another loon bobbed up in the water ahead, quite close to shore. It was another Arctic loon, swimming low in the water. It was moving towards the other bird swiftly, with a widening wake of silver in tow. We sloshed quickly along the spongy shore to the general area where we thought the second loon might have left and began searching carefully. If a nest was there, we felt that it had to be very close to the water, as loons are very awkward on land because their feet are placed so far back on the body.

In perhaps a quarter of an hour, we came upon the nest. Like that of the red-throat, it was on a mat of brown vegetation very close to the water's edge and contained the full complement of two large brown eggs. After photographing the nest, we moved off until we could barely see the general nesting site and watched for the better part of an hour. Both birds remained well out on the lake. Reluctant to disturb them further, we left.

On the ridges back by the canoe, several pairs of glaucous gulls were nesting. Each nest contained the usual three eggs, handsomely marked and well camouflaged. Having decided to try one of the pairs, we erected the blind at some distance. Mary saw me into the blind and then paddled away a goodly distance before beaching the canoe. The gulls would not realize that the blind was occupied, I felt certain. It would just be a question of a half-hour or so and I would have plenty of shots of them. After all, Bonaparte gulls, Franklins, ring-bills, Californias, glaucous-wingeds, kittiwakes and even herring gulls and greater black-backs had posed no problems whatever. All gulls are a piece of cake. Or so I thought.

The glaucous gulls were different. It was a full hour before one of the birds returned to a slight rise forty or fifty feet from the nest. I assumed that it was the female and carefully focused on the nest, hoping to get a flight shot of her landing. She failed to appear in my frame. Back at the rise upon which she had landed, she had begun to preen her plumage, running wing feather after wing feather slowly through her large yellow beak. She would finish any second, I thought, and then return to settle on her eggs. A half-hour passed. An hour. Two hours. No change. Still preening unconcernedly. Did she know that the blind was occupied? Was she aware of the fact that while two people had entered the blind, only one had left? After still another hour, the light seemed to fade strangely, but I had already christened her the champion preener of the entire Arctic and packed up my camera and lenses in preparation for leaving. Giving up for the time being at least, I signalled Mary.

Mary returned rather quickly, I thought. She came up to the back of the blind and began speaking anxiously. "I think that we had better get out of here, Cy. The fog is really rolling in off the ice pack. Didn't you notice the light changing? We have a long way to go."

I quickly crawled out of the blind and into a sunless world. Light fog was swirling all around, searching the area with grey, probing, amorphous fingers. The breeze was stronger now and directly out of the north where the ice pack lay. In that direction, the fog was an impenetrable moving wall, bearing down upon us inexorably and silently. Already the canoe, which Mary had beached a score of yards away, was little more than a hazy outline. We lowered the blind to half-height, snugged it down, stowed our cameras carefully and took our places in the canoe.

"I should have come sooner, I guess," Mary said. "But I thought that you would notice."

"Too busy watching the gull, but I should have been checking the light. The bright sunshine earlier fooled me. I should have remembered how quickly things

Short-eared owls could be easily spotted by their deep wingbeats. They nested on the tundra.

can change here with the ice pack so close. Would you mind straddling the bow, please? Watch the water in front of the canoe as closely as you can. If you think that we are straying from the channel, raise your right arm if you think I should move a bit that way, and the same with your left. I won't be able to tell from the stern. Let's go."

Mary moved to her position astride the bow, the motor coughed and caught, and we were off slowly. I strove to keep my ear tuned to the slowly turning motor so that I could tell when it was faltering in mud. Mary's body was bent sharply as she peered into the water below. Visibility was perhaps thirty or forty feet at most. It was imperative that we follow the narrow channel used on the way out. When Mary suddenly raised her right arm, I would attempt to compensate. Then her left, indicating that I had swung a bit too far.

Thus we made our way ever so slowly across the bay. I felt as though we were trying to navigate blindfolded in an eerie world totally devoid of reference points. I will never know just how Mary managed it all, but after what seemed an eternity, a bank loomed up dimly dead ahead and I shoved the motor into reverse. The canoe bumped, slithered and came to rest with front end raised as I quickly switched the motor into neutral. It seemed that we had reached the narrow channel on the south side of Fox Den Island. Mitten Bay, the broad bay that had concerned us most, must lie somewhere behind. We hoped that we had not followed the wrong channel and that we were, in fact, where we thought we were.

The fog was thicker than ever in the narrow channel, and visibility so limited that Mary, up front eighteen feet away, might have been a ghost from one of the Eskimo graves. Were we in the right channel? We should be breasting a light current but it was impossible to tell. An incoming tide could well be cancelling the current from upstream.

"I think that this is Fox Den, all right," said Mary. "The bank is quite steep."

"Then we must turn right. I think that we had better try for it. You never know how long this fog might persist and we don't have enough food to stay here for three or four days."

I pushed the canoe back into the channel with an oar, shifted into forward gear and turned right. We remembered that the channel that we wanted ran very close to the south end of Fox Den. Perhaps we could run close enough to keep bits of the dim bank in view. Mary signalled right and I followed.

"I can just make out a little of the bank at times and I'll try to keep you close," she advised. We crawled along at a snail's pace. The mudbars would not be a problem if we could maintain our position.

"The bank has disappeared altogether!" Mary sang out a few minutes later. "But I think that there is just a suggestion of current from the right now. Maybe this is the main channel of the river. What do you want to do now?"

"I'll try to hold her steady and if this is the main channel we ought to arrive at the cabin side before too long. Then we'll need to take a sharp left. Keep a lookout dead ahead and I'll cut the motor as soon as you see anything."

"I'll try, but this stuff is awfully thick."

We were travelling very slowly, reluctant to drive upon anything with force. As the minutes slipped by, I became convinced that we were probably crossing a broader span of water, which could hardly be anything but the main stream of the Anderson. "Any suggestion of a bank on either side?" I called.

"Not a thing!"

And then suddenly, the bow grated over pebbles. I flipped off the motor and quickly raised it clear. We had seen no gravel along any of the delta channels except the main river.

"Whew, I think we've made it!" Mary cried. "Talk about the blind leading the blind. I couldn't be sure of anything back there."

"Nor I. Let's finish that coffee and then we can take it easy back to the cabin. We can't go wrong now, even if we have to wade and pull the canoe. Boy, I'd hate to be lost in this stuff for several days. In the future, we had better take along more rations on trips like this."

We got back to home base with no more trouble and had a delicious supper of freeze-dried steaks, potato flakes, corn bread with syrup, and coffee. We celebrated with one of our few cans of peaches for dessert. The cabin was warm and snug with the stove blazing and the threatening fog, happily, on the outside.

Next morning, the sun was shining in a cloudless sky. The light southeast breeze had returned. We had just started in on a fine breakfast of toast, poached gull eggs, marmalade and coffee when a series of strident ptarmigan calls penetrated the cabin walls. They were like none of the typical calls that we had recorded and came in short bursts from the direction of Nest 2, behind and slightly north of the cabin.

We flung open the door and rushed around the corner. The male ptarmigan from Nest 2 was rising from the tundra near the nest on rapidly beating wings, calling wildly. Then he darted down at the willows, beating at them furiously and scolding at the top of his voice.

Realizing that something was very wrong, we ran forward to the nest. At that moment, the male stooped again, and then we saw the weasel. He was in the act of rushing at the incubating hen. She reared back on the nest and the weasel slipped his head beneath her breast and then backed away, clasping an egg between his jaws. Then he retreated to the cover of a denser willow thicket, apparently to consume the contents. The male continued the harassment, but to no avail.

"Bring my shotgun and birdshot, Mary. Quick!"

The weasel returned for another attempt, paying no attention whatsoever to my loud shouts. He made off with another egg. Mary arrived with the gun and I quickly loaded. In a moment, the weasel slipped forward yet again. The cock fluttered in the air above in preparation for another diving attack. I quickly aimed at the weasel, perhaps two feet from the hen, and squeezed the trigger. The weasel, arrested in midrush, slumped over motionless. Neither of the ptarmigan flushed.

With gentle fingers Mary slowly and carefully raised the hen from her nest. She had lost five of her eggs, but fortunately the last one laid and inscribed "10" remained with the other four. We needed that egg if we were to get a fairly accurate estimation of the incubation period. Then we scrounged among the willows, looking for the shells of the eggs. We found three large fragments, each containing the small perforations of the weasel's canine teeth, by means of which he had been able to carry off the smooth, rounded eggs.

On returning to the cabin, we finished our cold breakfast and I proceeded to skin the weasel and make it up into a study skin. As I was boiling the skull and cleaning it for preservation, I looked again at the needle-sharp canines. My memory stirred. We had lost a complete clutch of eggs at Nest 7, earlier. On checking a small bare area near that nest, we had come upon several fresh footprints of fox and little else. We had attributed the predation there to "probably fox." Were we right in our conclusion?

We had just witnessed ptarmigan nest predation by a weasel. Could the canines before me possibly assist in arriving at an accurate assessment of what had transpired at Nest 7?

We quickly dressed and headed north to Nest 7. It struck me that at the time of the loss of those eggs we had not thought of weasel, and even had the idea occurred to us, we would not have known where to look for evidence. Arriving at the site, we immediately dropped to our hands and knees and began to search carefully among the tangles of willow in the vicinity of the empty nest. At the end of about half an hour we had gathered no fewer than five large fragments of the shells of ptarmigan eggs. Three of these showed pairs of small, rounded perforations. We hurried back to the cabin and applied the canines in the jaws of the dead weasel to the perforations in the egg fragments that we had just found. The fit was perfect.

In our notes, we crossed out "probably fox" and substituted "definitely weasel."

Visitors

We were out on the tundra behind the cabin the following day when we heard the staccato "paw-ca-tah, paw-ca-tah, paw-ca-tah" sound of a helicopter overhead.

"Probably some oil exploration fellows on their way to work, away east of here. Seems to me that I heard that there is a crew in the Horton River area," I suggested.

The chopper broke through the thin layer of clouds and was clearly visible for a moment or two before vanishing again, heading eastward. We were surprised to hear the same sounds fifteen or twenty minutes later, this time from the direction in which the craft had vanished.

"That was a pretty quick trip," I said as the sounds grew steadily louder. "I wonder what's up."

"It's fog again," cried Mary. "Just look at it!"

I looked to the north and immediately became aware of a dense, greyish-white mass towering into the heavens. The breeze had shifted direction and was now blowing directly off the ice pack. Fox Den Island and Study Area had already disappeared into the advancing shroud. Then the indistinct form of the helicopter came into view, losing altitude and apparently heading for our tiny cabin. The forward line of the fog and chopper met on the gravel spit below, where we had pulled our canoe well above the tideline. Mary and I hurried down to meet our unexpected company. A smooth-shaven, parka-clad young man with an air of quiet confidence was the first to step down.

"Bill Armstrong, Penguin Petroleum," he said, extending his hand and grinning widely. "We were headed for the Horton but got socked in like you wouldn't believe! Saw your quarters a bit before we hit the fog and hoped there'd be somebody home."

"Cy and Mary Hampson, Edmonton. We hadn't noticed the fog until we saw you circling back."

"Say, could you put us up for a bit? Just until this damned stuff lifts?"

"You bet. We'll be happy to do whatever we can."

"But I've got quite a crew in there," he said, jerking his thumb in the direction of the sizeable chopper, a big Sikorsky.

"We can always squeeze in a couple more." I hadn't thought of eleven more.

"Like to start with coffee?" Mary said as we watched the crew climb out. Ten of them!

"Tuffy won't be interested in coffee right now," Bill explained, with that same likable grin. "He took on a few too many in Inuvik. He's sawin' California redwoods with knots in them a foot through by the sound of 'im. Maybe he'll make it later if he can find his way to your cabin in this bloody fog."

We climbed the bank to the cabin, following the narrow trail in single file. What were we going to feed them all? How come I'd completely forgotten to bring along the fast-food hamburger house on the highway south of Edmonton?

"You'd better make the coffee in that big open kettle, Cy," Mary said in her cheerful business-as-usual tone. "I'll mix up a load of biscuits and pop them into the drum oven."

Our guests arranged themselves on our four chairs, a wooden bench and along one side of the bunk. Two of them could find no other place, so sat on the floor with their backs against the wall.

"Sure snug in here," said Bill enthusiastically. "We ate in Inuvik. But I hope for your sake," he continued, turning to Mary, "that the soup out there lifts sometime this spring."

"We're really happy that you dropped in," replied Mary convincingly. She has always been a people person with a way of taking things in her stride. "You really did drop in, didn't you?"

"That's closer than you think. And thanks very much. I don't know where else I could have set her down with any chance of getting through to Inuvik if we have to. We didn't get much notice and the visibility ahead was about minus twenty! But mebbe I should introduce these mugs to you, eh?"

"Good idea," Mary agreed with a smile. "It's not easy to carry on a conversation when you have to say: 'You, over by the door, or 'You, with the big smile.'"

"Right. Well, all of these mugs ought to, by rights, be in Alcatraz, but for some reason or other, they chose the Arctic instead. Don't ask me why. That long geared bird on the end is Gene. He's a whale of a catskinner. Like Tuffy down in the chopper, but Gene holds his liquor better. The two fellahs next to Gene, we call Pete and Repeat. Both oil geologists and if there's any up here they'll find it. They got into a bit of overproof oil in Inuvik, too, but no harm done. The two jokers on the floor are Jim and Holy Moses. They're first-class tool pushes for sure, but they play poker even better. The real handsome guy with the cookie duster is Jake. He'd like to get back to Hollywood. Says the gals up here wear too darn many clothes.

"Billy and Gord there are the best mechanics this side of the Arctic Circle. Both Eskimo, and they sure know how to look after a rig! Ralph, over on the far end of the bunk, is a whale of a second pilot. He's not flyin' today. He likes it when the going is tough. You oughta see him handle one of these kites in a blinding snowstorm! Last of all, is that pint-sized lug next to Ralph. That's Randy and all I can say is that he's got to be mebbe the best damn whirlybird pilot God ever created. Nobody should get in a fight with the little squirt, though. He's got two fists on the enda each arm and you never know which one's comin' at you. Fellow in Inuvik took him on but I don't think his mother'd recognize him today."

"Thanks, Bill. It's nice to meet your crew."

Everybody seemed to enjoy the hot biscuits smothered in melted honey. At any rate, they vanished faster than the proverbial snowball. So did the coffee. Bill walked over sprawled legs to the window and peered out.

"Doesn't look like she's lifting," he announced. "And darn near time to turn in. The fellows didn't get much sleep last night. Okay if some of us bed down here, Cy?"

"Use any space that you can find. You won't bother us."

"Say, you didn't happen to hear a whirlybird going over here last Wednesday, did you, Cy?"

"No, yours was the first."

"Well, they mighta gone over further upstream and missed you. We had a bad accident with one of them back by Eskimo Lakes," he went on, jerking his thumb to the west. His voice seemed tinged with concern and his usual grin was absent.

"What happened, Bill?"

"They landed on the frozen lake for some reason. The tracks are all there. Then they radioed that they were taking off for Inuvik." Bill hesitated, his eyes looking into space, before continuing. "But they didn't get to Inuvik. We found the outfit about a mile down the lake from where they had landed. She was all piled up and all four of 'em were killed."

"Man, that's a tough break. Any idea what caused it?"

"We think so. The propeller was some distance from the crash. One of our mechanics took a good look at the wreck and the snow around. He figures that they took off okay and got up some distance before the shaft broke. She must have come down like a lead balloon. He says that the shaft looks like it crystallized, you know, the way steel does sometimes."

Gene had evidently been listening. He broke in, "They never had a chance. They musta come straight down. Not a chance!"

Bill seemed anxious to change the subject. "All right, fellows, it's after midnight. Time to get some shuteye. Gene can sleep in the bird with Tuffy, but the rest of you had better get your sleeping bags and bring them along here. Would there be room for two in the shed behind, Cy?"

"Sure, they can shove the gear in there against the wall. I'll light a bit of a fire in the stove to take the chill off. There's some foam rubber in there that they could roll out their sleeping bags on, too."

Minutes later, when we had all flattened out, the small cabin was bursting. One was stretched out on the narrow bench, two were under the table, one back of the water drum and two alongside the stove. We finally persuaded Bill to unroll his bag on the bunk beside us. That was all the space there was left.

The snoring chorus was more varied and higher in amplitude than anything we had heard from the birds on the open tundra, but I think that we all drifted off in time. It seemed only a few minutes later that Bill unzipped his sleeping bag, but looking at my watch, the hands stood at 6:30. Bill was peering out of the window.

"She's all gone, fellows!" Bill called. "Daylight in the swamp. We've got work to do!" The crew rolled out one after another, rubbing their eyes with their knuckles.

"When's a fellah going to get a chance to sleep on this job, Strawbones?" growled Jake, the Hollywood type.

"As soon as you get back to Edmonton on your week off. Three more weeks, Jake, and then you can sleep around the clock for seven whole days."

"But that's girly week!" advised Jake, challengingly. "I never sleep then."

"Cut it, you're bragging again, Jake. You and Gene hustle down to the bird and bring up some stuff for breakfast. Don't forget the bacon and eggs. Better bring Tuffy, too. He'll be as hungry as a bear after sleeping all winter."

The fog had disappeared, all right. When I went out for some wood, the sun was beaming down. The wind had changed to the south again and only a single cumulus was coasting serenely northward on the eastern horizon. Jake and Gene were back shortly with their arms full of cartons. Tuffy slogged slowly behind, but he was steady on his feet.

"Bacon, eggs, bread, marmalade, strawberry jam, canned milk and coffee," sang out Jake. "Anything else, Chief?"

"That's about it, Jake. Did you bring lots?"

"Damn right! Sorry, Mary. Darn right. Everybody's hungry."

"Bill, maybe you'd like to help Mary with the bacon and eggs on the gas stove. I'll brew up the coffee on the other burner and maybe two of your men could look after the toast on the heater. It works pretty well when it's hot. Okay?"

"Yes, sir!" said Bill with his usual grin. "Just tell me what to do, Mary. Gene, you and Jake can do the toast. Ralph'll butter it."

"We'll have to eat the bacon and eggs in shifts, maybe," I suggested. "We don't have that much cutlery or that many plates."

"Or some of you could use paper towels if you like and put your eggs right on your toast," Mary offered. "We had enough cups for last night, so we should be all right in that department."

The coffee was soon boiling. Mary was laying the bacon out in two pans, while Bill turned it and piled it on a tin plate on the back of the heater when it was done. The pile of toast grew at first, but as soon as some of the bacon and eggs were ready, attrition set in and Gene and Jake struggled to keep up. They had brought in a small mountain of grub. Half a dozen loaves of bread, large tins of jam and marmalade, two pounds of coffee, three or four pounds of butter, four dozen eggs and enough bacon to feed a threshing crew on the prairie. But at the rate it was disappearing, there wouldn't be much left for the bears.

"Anybody want more toast?" piped Gene, after most of the fellows had slowed down.

"Two here," said Tuffy, the catskinner.

"One more here." This from Pete.

"Okay, time for a couple of you well-fed steers to take over here," announced Jake. "You don't know how hungry Gene and I are! We've been smelling that stuff for an hour and our stomachs are yelling that our throats are cut or plugged with drilling bits."

Ralph, Bill, Gene and Jake took over at the table while others took their places. Mary had quickly washed four of the plates, forks and knives.

"How about you and Cy?" Bill pressed.

"We've got all day," Mary replied. "I'm getting a real kick out of watching an oil crew eat." At length, everybody seemed content foodwise, and they began to leave.

"Thanks, Mary and Cy. It was awfully good of you to share your digs with us."

"You couldn't be more welcome," Mary smiled in return.

Everybody had left but Bill and Jake. Half-turning to leave, Jake took Mary's hand in his rough one. "You know, Mrs. Hampson . . . uh . . . uh . . . Mary, you're in the wrong place for sure."

"How so, Jake?"

"You'd really go over in Hollywood. Big!"

"Thanks, Jake," Mary smiled warmly. "But I don't think that M.G.M. would have room for Cy and all his ptarmigan."

Jake left and Bill extended his hand. "A million thanks to both of you. Your warm hospitality is something I'll always remember. You put up quite a crew here."

"It was fun," Mary smiled.

"Look, we'll likely be out this way again. Two or three of us, anyway. Is there anything that we could bring you from Inuvik?"

"Thanks, Bill. If you happen to run across a bit of fresh fruit or a head of lettuce or cabbage, that would be great, but we're fine. Dinna fash yersel'," she concluded, her Scottish ancestry coming through. "Oh, by the way Bill, you sure know how to handle your crew."

"They'd be hard to beat, for sure."

And he was gone. We watched his tall form retreat below the riverbank and waited for the chopper to start. We finally heard the big Sikorsky cough and begin to warm up. Then Jake suddenly appeared, running towards us over the lip of the bank. He was carrying a large carton. He set it down in front of Mary and stood for a moment, breathing hard.

"Bill said he thought you might be able to use some of this stuff. We haven't got any room in the bird for it."

"Thank him for us, Jake," Mary responded, with an understanding grin. "It has been a nice break meeting the crew."

Jake started down at a trot and then half-turned, waving a hand over his shoulder. "See you in Hollywood!" he flung out.

Mary stooped over and pushed aside the loose flaps of the carton. "Bacon, two dozen eggs, real butter, coffee, marmalade, five bananas, two cans of peaches, two cans of cream and four loaves of bread. Who needs M.G.M.? I don't have to bake today."

Her high spirits were infectious. "And since you're not going to California for an audition, maybe we can adjudicate the performances of some of the players on the tundra stage right here. Maybe one of them will win an Oscar. Who knows?"

"They were certainly an interesting crew," Mary said. "I like the way they pitched in."

"Me, too. But I still have one very big concern."

"What's that?"

"I just wonder what impact their industry will have on this very fragile environment over the next few decades. If they go ahead with more of those manmade islands offshore for drilling and have blowouts or accidents such as have occurred in the Gulf of Mexico, Valdez and the Strait of Hormuz, heaven only knows what will happen to the beluga whales, seabirds, waterfowl and other stuff. This is the last frontier, and I don't know how an acceptable cleanup could be organized and carried out in this very vulnerable environment."

We did up the dishes, straightened things around a bit and then spent the rest of the day checking our ptarmigan nests. There was still no sign of hatching. The small cabin seemed four times its usual size when we turned in that night.

Again, we awakened to a cloudless morning with a soft breeze from the southeast. Along with our cameras, we put extra rations into the canoe and headed again for Moose Track. The two pairs of striking loons and the glaucous gulls which we had found there demanded attention. Remembering the problem of the fog on our last trip, we paid particular attention to the channels on the way out in the hope that we would be better able to remember them should we be caught in another fog.

Arriving at Moose Track, we first checked the blind that we had left at some distance from the glaucous gull and then put up the second blind near the red-throat nest to give the birds a chance to become accustomed to its presence. They did not seem particularly alarmed and remained near the centre of the small lake, rather than retreating to the far side. That was encouraging. Then another try at the glaucous gull.

She was still absent when we returned from the red-throat site, but I crawled into the blind nevertheless. Mary left with the canoe as before, and I had a lengthy encore, with the gull perched on the ridge some distance from the nest. There she preened and preened as though preparing herself for a command performance with the queen or an audition with Twentieth Century Fox. Speaking of vanity! I was totally frustrated after two hours of the performance and signalled Mary.

"I'd like to try her," Mary volunteered. "You never know, maybe she prefers girls. She certainly doesn't have much patience with the opposite sex. In humans, anyway. Maybe you're just losing that Don Juan touch."

"Crawl in, then. I won't be jealous if she finds you a kindred spirit. Good luck!"

So we exchanged places and I pulled away in the canoe, leaving Mary in the blind. Fat chance, she had, I thought while nosing the canoe into the bank where Mary had parked it. When I glanced back, the gull was on the nest. Just like that. Fifteen minutes or so later, Mary draped the signal ribbon out through a hole in the blind. I backed out, started the motor and returned to the blind. The gull flew off lazily.

"Absolutely nothing to it," Mary laughed. "She's a real sweetheart. Would you like to try? Now that I've won her over?" she added with a broad grin.

"You bet your wedding slippers, I would! Did you get some good shots?"

"Of course. I enjoy working with very cooperative birds. I'll take the canoe to

Glaucous gulls supplied us with fresh eggs early in the season.

the same place. You'll be finished in a matter of minutes. Now that I've got her reconciled to our presence."

Reconciled? Cooperative? A real sweetheart? Nothing to it? The gull returned to preen for another two hours in the same spot as before. I signalled Mary, moved the blind to about twelve feet from the nest and left the gull to her. I hadn't gone halfway to the beaching point before the gull was back on her eggs. No loafing around. No tantalizing. No preening. What did she have against me, anyway? Only then did it strike me. Mary's eyes were more hazel-grey than mine!

We tried the red-throated loon next. She came in quickly when the blind was a little more than a dozen feet away, so I decided to move it still closer in the interests of capturing even more detail. About ten minutes after Mary had left, I peeped out cautiously. One of the pair was swimming towards the nest. It suddenly dived, slipping beneath the surface with hardly a ripple. When it surfaced again, the loon was close in and directly in front of the nest. Reaching the edge of the water, it slithered up the far side of the nesting mound on its belly and in a moment stood vertically over the eggs, turning its head from side to side while peering down at the eggs. Only then did I fully appreciate how far back on the body the legs are placed, and why they have so much trouble travelling out of their element, the water.

The loon then bent her neck sharply downward and deftly turned the eggs with her beak, tucking them against her large webbed feet. Then she lowered her body carefully, wriggled into a comfortable position and settled to incubate, broadside to the camera. I caught my breath, overcome with the magnificent sight before me. The grey head and side neck were smooth as silk but glowed with the texture of velvet. The piercing red eye complemented the rich red throat, while the white upper breast shaded gently off into the lustrous grey back. My eye traced the fine, matchless lines running up the back of the neck and then leaped to the ebony beak, tipped ever so slightly upward.

When I had finished, Mary went into the blind for a spell. It would be an experience that she would never forget. She emerged fifteen or twenty minutes later, her eyes shining.

"What a bird! I wouldn't have guessed the effect of those colours and trim lines when seen so close. She's absolutely superb! She even called to her mate on the water and he answered. Did she do that for you?"

"No. How did you like her eyes?"

"Living jewels! Especially with the warm light in them."

"Let's try the Arctic loons on the other lake. They're colourful, too."

Through with the blind at that location for the time being, we took it down, rolled it up and started off for the larger lake where we had found the other loons. We were startled when a medium-sized brown bird suddenly flushed at our very feet and took to the air, calling loudly. It landed a short distance in front of us, with wings raised, and continued calling as though in distress. We immediately noticed the very long, down-curved beak.

"Hudsonian curlew! Whimbrel! How lucky can we be? The nest must be right here. Watch your feet!"

We stood in our tracks and searched the grass-grown bit of tundra ahead of our toes.

"I see it," Mary said. "Four beautiful eggs."

The nest consisted of a shallow grass-lined pocket, hidden only by the grass growing around it. The four large eggs filled it to capacity and, as in nearly all shorebirds, their blotched colouration provided effective camouflage. The curlew was still striding about, calling at intervals. I photographed the nest and then decided to try for the bird. We moved back a short distance, set up the blind and Mary walked off with her field glasses slung around her neck.

The curlew couldn't have been more cooperative. She ceased calling and in minutes I could see her striding on long legs towards the nest. She paused briefly at intervals, often in midstride, cocking her head at the sky above. Then on she came. Arriving at the nest, she carefully placed a foot on either side, fluffed out the feathers on the lower breast and gently lowered herself onto the eggs. Then a little shimmying, and she was still.

A most striking bird in her well-tailored, soft brown plumage. Her curved beak, though not as long as that of the long-billed curlew of the southern prairies, was impressive and well adapted to picking insects and spiders from low-growing vegetation. White throat, greyish line along the crown, dark stripe through the eye and the lines of dark spots down the breast all combined in creating an image in muted, blending colours.

Then on to the Arctic loon. Here, unlike the red-throated loons, both adults

Whimbrels or Hudsonian curlews, probably from Chile or Brazil, nested on the tundra.

were resting in the water near the far side of the lake, apparently unhappy about our being in the vicinity of the nest. In our field glasses, they appeared alert and apprehensive. We decided to begin well back, since it looked as though they might not accept the blind at close range. As it turned out, we had guessed correctly. It was only after twelve days of repeated attempts that one of the loons plucked up sufficient courage to return to the nest with the blind in place. We were reluctant to keep them off the nest for lengthy periods, fearing that the eggs might become chilled. So our trials were short in duration and after each, we moved the blind far back in order to allow the loons to resume incubation.

However, on the thirteenth day there was a change in behaviour on the part of the loons. In less than a half-hour, one of them swam directly to the nest and immediately settled over the eggs. And what an incredible picture she made. Her glowing pearl-grey head and sharply checkered back, together with the wine-red eye, brilliant neck stripes and sharply contrasting jet-black throat and immaculate breast, created a veritable masterpiece in line, form and texture.

Male willow ptarmigan in full breeding plumage.

A pair of peregrine falcons nested on the Anderson cliff.

Arctic tern. The flight pattern was irresistible.

A pair of tundra swans at their nest.

Mary with day-old tundra swan cygnets.

The Arctic loon was a superb study in black and white.

Red-throated loons nested on the smaller tundra lakes.

The midnight sun casts a golden glow over the landscape.

Lesser snow geese were common breeders on the Anderson Delta.

A white-fronted goose returns to its nest.

Incubating female willow ptarmigan were wonderfully camouflaged.

Willow ptarmigan attended their new chicks closely.

Mary checks Arctic tern's nest; Arctic tern checks Mary.

Arctic ground squirrels occasionally came for a handout.

Red-throated loon deftly turning her eggs before settling.

Semipalmated plover nested within the territory of an Arctic tern.

Old squaw ducks often nested within the territory of an Arctic tern.

Incubating male northern phalarope often shared a tern's territory.

Snow geese on the Anderson Delta.

American widgeons consorted with tundra swans.

Common snipes bred in the low wetlands.

Mosses flourished in wet areas during the brief summer.

"A wolf at the door."

White-crowned sparrows were common on the delta.

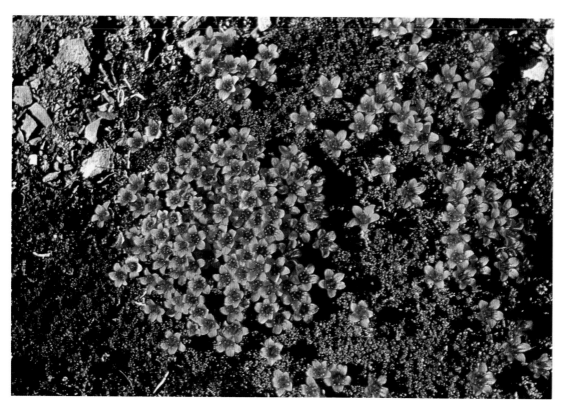

Purple saxifrage carpeted much of Bathurst Island.

Pedicularis arctica, Anderson Delta.

Sanderling chicks are studies in consummate camouflage.

Red phalaropes can be difficult subjects to photograph.

Camouflaged king eider females require little cover when nesting.

Rock ptarmigan and chicks withstand the bitter cold of the High Arctic.

Newly born Arctic hares blend with the brown tundra.

Handsome downy chicks of the black-bellied plover

New Beginnings

By mid-June and early July, new additions to the populations of arctic breeding species were occurring daily. Not the least of these concerned the veritable explosion in the numbers of the ubiquitous mosquito. They swarmed about our heads in their humming hordes and drank incessantly of the blood from beneath exposed areas on hands, neck, forehead and cheeks, while the constant drone of their millions could only be escaped by remaining indoors, by building densely smoking smudges on the riverbank, or by that great but undependable gift of the arctic gods, wind.

Since remaining indoors was out of question except for the writing of notes, breakfast and the last meal of the day, we fought mosquitoes on their terms. In the field we attempted to counter their sheer numbers with generous applications of a recommended insect repellent, and by wearing the rubberized nylon pants and jackets that we had found so useful earlier, when cold winds were relentlessly knifing across the tundra or when icy spray whipped over us from the bow of the speeding canoe. But the trouble now was that while the almost impenetrable clothing assisted with the problem of mosquitoes, it also tended to retain body heat and contributed greatly to physical discomfort.

Caribou were also plagued by these stinging, bloodsucking legions. They responded by trotting out to the Arctic coast, where breezes off the ice pack sharply reduced the numbers of the insect pests attempting to harvest from them. The scourge was at its peak on still days, particularly when the sun dipped to its lower points above the horizon.

It was truly surprising how numerous these little blighters could be. On one occasion when their clouds were unusually bothersome, I bared the back of my hand to them for ten seconds, allowing them to gather. Then I swatted swiftly with my other hand to see how many of them I could kill. The mosquito corpses totalled eighty-three!

The eggs of the geese began to hatch first. The golden-yellow goslings of the lesser snow geese were in marked contrast to the dark grey, downy young of the brant. The goslings were barely dry when both adults conducted them to the nearest water. Nevertheless, this was a critical time in the life cycle of these birds, since there were many predators eager to take advantage of the new supply of tender flesh. On one occasion, a large, very dark wolf raided one of the colonies and made off with a number of them, despite the fact that the adults protested vigorously with flapping wings and high-pitched scolding. The wolf's strategy was to swallow as many as possible before leaving. Such quantities of food would likely be then transported in his stomach back to the litter and regurgitated for them at the den.

Light, soft-stepping foxes took their toll, too, along with parasitic jaegers and foraging glaucous gulls. Both gander and goose tended to martial their goslings closely between them as they made their way as quickly as possible to the water. The swiftly stooping jaegers picked off their share of the harvest despite the efforts of goose and gander to beat them off. The slower, heavier glaucous gulls continued their efforts even after the goslings had reached the water, but at this point the jaegers discontinued their direct raiding tactics on the broods.

Instead, they flew vigilantly overhead or watched the big gulls from the vantage point of a slight rise in the terrain. Whenever a gull successfully picked off a gosling and rose into the air to make off with the chick dangling from its beak, the jaegers began to divebomb the gull and harass it mercilessly. In its efforts to escape its pursuers, the gull often dropped the chick, whereat one of the jaegers stooped swiftly and caught it in midair. They were living up to the reputation signified by their common name, "parasitic jaegers."

The slim, graceful Arctic terns with their long high-speed wings and flexible swallow-like tails were the most successful birds on the delta when it came to protecting their eggs and chicks against predators. They flew swiftly out to meet potential enemies whenever they entered their territories and proceeded to divebomb them persistently. When we went over to check their eggs, we learned just how serious the terns were in their defiance of potential predators. Yet their matchless flight patterns were so incredibly attractive that one had to attempt to capture some of them on film.

Accordingly, I quickly mounted a short focal length lens on my camera and aimed it at a tern diving vertically at me. I released the shutter and almost on the instant felt a sharp stinging sensation in my trigger finger. Glancing down, I noticed a trickle of blood running across my knuckle. The tern had stabbed me with its needle-sharp beak.

"Ouch!" cried Mary at my side. "One of them got me squarely on the top of my head."

"Better cover it," I advised. "These birds mean business."

Mary pulled her parka hood securely over her head but the terns kept diving. I wanted to try for a high-speed shot of one of them attacking her, so I focused sharply on her head. One of the birds dived swiftly and landed upon her hood with wings extended and then paused briefly before pecking fiercely at the protective covering over her head. I released the shutter.

"The fanciest head gear I have ever seen," I offered. "Better than a bouquet of roses, carnations and orchids, any day."

"These terns neither feel nor act like carnations and orchids, believe me!" Mary had stretched out on the tundra, her chin in her hands, as she carefully checked the tern's eggs for signs of hatching. The eggs lay on the open ground about a foot in front of her.

"Hey, the nest here has only two eggs in it. Not like the nests of terns back

Arctic terns, having returned from regions in the Antarctic, nested on the delta.

on the prairie. Black terns, commons and Forsters normally lay three. I don't think that we have seen a single Arctic tern's nest with three eggs, have we?"

"No, that's right. And that must mean that Arctic terns have a higher rate of survival than black terns, common terns or Forsters. Or else their life spans are longer. I read somewhere that Arctic terns have been known to live for twenty-seven years."

"That seems quite an age for a bird with such a long and hazardous migration," Mary observed. "And I would bet that the chicks have a very high rate of survival. Mighty few predators would succeed in snatching either eggs or chicks, considering the determination of the adults to guard them," she continued. "By the way, both of these eggs are pipped."

As I started over to look at the eggs, one of the terns landed lightly upon her head and perched there, quietly surveying her eggs in front of Mary's nose. It appeared that they had either flown out of steam or decided that we did not pose a threat. On getting up, Mary flushed a northern phalarope from his nest only a few feet from that of the tern's eggs. And as we turned to leave, we spotted an old squaw duck incubating her clutch of eggs, very close to the nest of the phalarope. We later noted that this was a common practice with these three species on the south end of Study Area.

Neither the old squaw ducks nor the northern phalaropes attempted to challenge potential predators when they chanced by. Rather, they appeared to take advantage of the fierce defence put up by their alert feathered watchdogs, the Arctic terns. It would be interesting indeed to know how this advantageous behaviour pattern evolved.

The very diminutive semipalmated plover were nesting in the gravel, a short distance up from the river's edge. They were almost the colour of the common killdeer, but half their size. The snow-white breast was crossed by a single black bar rather than the double one of the larger plover. Their eggs had been difficult to find because of their consummate camouflage. The newly hatched chicks, spotted and blotched to blend with the sand and pebbles surrounding them, were almost invisible. We photographed the adults at very close range, using a blind to secure the desired detail in their brilliant plumage. The bright golden eyering of this small plover would be the envy of any human bride, though of course she'd prefer it on her ring finger.

The very small semipalmated sandpipers, which had earlier serenaded us throughout the day during courtship, had chosen to nest on higher ground, hiding their nests among the grasses and low-growing willows. On checking one of their nests on the day following our work with the plovers, we found that all four eggs had hatched and the chicks had almost completely dried off. After getting my camera from my pack and fitting it with a macro lens for closeup work, I knelt down on one knee to photograph the downy brood. The brim of my hat was casting a shadow over the youngsters, so I laid it aside on the tundra. But before I could bring them into sharp focus, they began to scramble out of the nest.

This wouldn't do. The chicks would scatter and the adults might have a good deal of trouble rounding them up. I grabbed my hat while Mary quickly scooped up the chicks and deposited all four of them in it.

"Just set the hat down for a second while I get a shot of them. Then we'll settle them down together in the nest and clear out in a hurry. She will pick them up before they can scatter."

Just as I leaned forward to photograph the chicks, the mother suddenly appeared from the surrounding grass, calling in light plaintive tones. She settled on the brim of my hat within inches of my face and continued with her anxious notes. The chicks began peeping in return and in a moment one of them had scrambled out of the hat. The mother immediately fluffed out the feathers on her breast and the chick wriggled below them. The others followed at intervals and before long the mother was brooding all four of them on the brim of my hat. At length, having warmed her very young brood, the female led off and her chicks followed.

Semipalmated sandpipers had never been easily approached as they busily foraged along the shores of lakes and oceans during migration. Yet this adult had settled unhesitatingly upon the brim of my hat, barely inches from us both. She must have experienced an inner conflict in which she was torn between two drives, the urge to flee and the urge to care for her chicks. As in the tern, the urge to minister to the needs of her offspring had been the stronger.

The young ground squirrels had now appeared above ground. They seemed to be continually engaged in feeding. They grew rapidly and began to put on fat in preparation for the very long period of hibernation ahead. It did not seem possible that they could be ready when the time came in the very near future, when the tundra would once again be blanketed with snow and howling arctic gales would drive it into deep protective drifts along the riverbank.

The willows had long since shed their pussies and many flowers were in full bloom. Dryas, or northern avens, the floral emblem of the Northwest Territories, carpeted much of the tundra with their bright white faces and centres of brilliant sunshine. Scarlet Arctic rhododendron bloomed in the damper spots among the willows while waxy, yellow marsh marigold and handsome clusters of Labrador tea flourished in lower areas with their roots in shallow water. Striking, down-clad woolly lousewort extended its compact, bright red flower heads towards the sun, while its close relative, the Arctic lousewort, was resplendent in shades of pink. But the most impressive member of this tribe was *Pedicularis capitata* with its large, lustrous, showy flowers, their glowing surfaces reminding us of the sheen of shot silk.

Purplish-red Indian paintbrush reared its proud head among the dried stalks of the preceding year's growth. Several species of slender wintergreen added their bright flowers to the variegated greens of mosses and the golds of lichens. Tall cerulean lupines bloomed in incredible abundance; the fragrant aroma borne on breezes blowing over millions of acres of them was soul-searching in its intensity.

It was time to check all of our ptarmigan nests and score the results. This proved an exciting business, especially since we had come to know many of them individually, not only through the individual territories that they occupied, but also as a result of minor variations in both plumage and behaviour. A number of idiosyncrasies had been recorded as well. In the Don Juan pair, the male had on several occasions been observed attempting to court the hens of other established cocks. The two Bear Flag Gals shared a male and the Courting Couple had spent an inordinate amount of time in courtship, never seeming to surfeit during that stage.

White-Rump retained a patch of white feathers in the rump area after moulting into the brown summer plumage; the combs of Pinky were always coral-pink in colour, rather than bright scarlet as in other cocks. Dark Boy, partially melanistic, was decidedly several shades darker than his male associates. Dumpy had attempted to lay two eggs in her nest during the severe storm but had, on both occasions, "dumped" them some three feet distant. The Paddle Pair had chosen a nesting site close to an Eskimo grave on which a paddle had been left at the time of the burial of the occupant. The Tall Willows couple had hidden their nest at the base of the tallest clump of willows in the area, while the Bluff Willows pair nested in willows at the foot of South Bluffs. On our map we carefully designated all of the nesting sites by their coordinates on our map: 0,50; 650,25; 1300,300; etc.

On checking 0,50 we found the hen broadly flattened on the nest. She was reluctant to leave, even when Mary stroked her gently down the back and wings. On raising her carefully from the nest, we discovered that all of her eggs had hatched and the chicks had already dried off. Ten unbelievably handsome chicks clothed in silky down in shades of brown, buff and gold. Each sported a chestnut-red baseball cap and a tiny ivory egg tooth near the tip of the beak. The hen remained on her belly beside the nest for several minutes, clucking softly at intervals. Then she slipped back onto the nest, tucking the youngsters from view beneath her breast.

The eggs had all hatched at Nest 13 as well. The hen flushed when Mary stroked her back, and exploded into the air. She flew at Mary, striking her with wings and scolding hoarsely. Then she settled on the tundra a few feet away with her wings broadly spread, chirring and scolding anxiously. We colour-marked each chick on the left leg with a felt pen, recorded the pertinent data and left. The hen immediately returned to her chicks. And so with the remaining ptarmigan nests.

A good deal of interesting information was gathered in connection with the ptarmigan study. The territories of breeding pairs were quite distinct. Any given nesting site was, on the average, just over a hundred yards distant from the next closest one. Each territory contained a dust bath, which was used frequently by both birds, and each pair tended to use a specific feeding area located outside the nesting territory. Egg-laying began during the last few days of May, with additional eggs added at intervals of twenty-four to thirty hours.

The cocks conducted the hens to the nests for egg-laying purposes and stood guard nearby while they were on the nests. At such times the hens remained on the nests for about two-and-a-half-hour periods. They usually covered the eggs with wisps of grass, moss, lichens and leaves on leaving. The hens laid 7.6 eggs, on the average.

The hen carried out all of the incubation duties, while the cock stood guard until she left to feed. The cock followed, again remaining on the alert while she fed hurriedly. Feeding periods were short, generally lasting eight to ten minutes. As incubation proceeded, the cock took up his station closer and closer to his sitting mate. On one occasion, when we flushed a hen in order to check her eggs, she ran directly towards the cock, who responded immediately by running towards the approaching hen. While passing her, he fanned his tail and fluffed out his neck feathers. The hen continued on for perhaps twenty-five feet, churring and clucking, and then settled on the tundra on her belly.

The cock continued on towards the nest, bringing up a short three feet from nest and observers. Here, he drooped his wings, fluffed out his feathers, raised his head and began to hiss vigorously. As we departed, he raised his voice in the characteristic "G-dout! G-dout!" warning.

The incubation period was very close to twenty-one days, the same as for domestic chickens. 73.1 percent of the eggs originally incubated hatched successfully. Most of those lost were due to predation by weasels. Egg fertility in all of the nests not raided by weasels totalled 96 percent . The chicks left the nest within a few hours of hatching, almost always attended by both adults. Their camouflage was so effective that they were almost impossible to see until they moved.

Perhaps the most interesting aspect of this study lay in the remarkable synchronization of events in the breeding cycle of these fascinating birds. All of the eggs which hatched in all of the nests under study did so in the short time-span of five days. With such very short reproductive seasons in regions well inside

Willow ptarmigan chicks in nest.

the Arctic Circle, there is little time to waste. The ptarmigan had arrived on the breeding ground in late winter and had established their territories and begun vigorous courtship activity while the tundra was still heavily blanketed in snow. Even mating and egg-laying had proceeded in late-winter gales. Incubation followed closely on the heels of the laying of the final egg, and in three weeks the chicks saw the light of the long arctic day.

The newly hatched chicks had begun to forage energetically for food as soon as they had dried off and, though still very small and ill developed, were able to fly well by the fourteenth day. Even this barely gave them sufficient time in which to mature to the point at which they would be able to migrate some distance south and endure the rigours of the ensuing subarctic winter there. Fortunately, the early days of the chicks were twenty-four hours in length, which enabled them to feed and develop around the clock.

In retrospect, it does not seem possible to overemphasize the importance of the behaviourial patterns of the male willow ptarmigan in the survival of the hen and the generation that she had hatched. He had been constantly on guard throughout the incubation period, warning his mate whenever potential predators appeared, and this concern continued as he followed her and the brood about during its period of development. Successful reproduction is critical in the survival of any species.

At the end of the season, we left the Anderson delta with deep regret. As our plane circled to gain altitude, we looked down upon the tiny red-roofed cabin which had served us so well, then struck off over rivers, green-stippled tundra and multicoloured, iron-charged lakes for Inuvik on the mighty Mackenzie.

We had had the transfusion we sought.

To Bathurst Island

I vividly remember a telephone call, way back in 1968, which set in motion a chain of events that will remain indelibly stamped in my memory so long as I am able to draw a breath.

"Hello."

"Hello. Ottawa calling. MacDonald, National Museum. That you, Cy?"

"Sure is. How are you, Stu?"

"Just fine, thanks. Busy as ever, though."

"It's good to hear your voice. What's up?"

"I've been wondering if you'd like to join our party this spring. Thought I'd better check with you."

"Where are you heading? Coming out West again to work on grouse?"

"No, we're going into the High Arctic to explore the last of the unexplored High Arctic islands, Bathurst Island. It has been surveyed but never explored, neither zoologically nor botanically. We'd like to know what's there."

"Man, that sounds exciting! But when are you going? I'm not sure that I can get away. It depends upon how early you want to leave."

"Well, we'd like to be there for the last week in May. We could set up camp and you could come along whenever you can make it. How does that sound?"

"It sounds great and there is a good chance that I can get away by then. I really appreciate your calling, Stu. I'll see what arrangements I can make here and keep in touch. I'll call you as soon as I can."

"You do that. Say, we're going in from Montreal, but it might be a lot easier for you if you flew directly from Edmonton. Let me know so that we can pick you up at the other end."

"Thanks a million, Stu. I can't think of anything I'd rather do."

"I hope you are able to make it. All the best to everybody there."

Two months later, at night on May 27, I boarded a huge Hercules transport in Edmonton, Alberta, with a load of propane and an oil crew of sixteen bound for Melville Island off the Central Arctic coast. It was raining hard when we took off but the aircraft soon broke through the heavy layers of dark cloud into light fleecy skeins. At eighteen thousand feet, a narrow strip of rose tinged the sky to the northeast and the ripples of cloud below reminded me of the ripples on the pale windswept Sahara.

We were in the belly of the big transport, the centre occupied by two long rows of propane tanks held down with a large sheet of perforated plywood. Through each hole in the plywood protruded the neck of a cylindrical propane tank. The stout wooden sheet supporting the tanks was securely roped to eyebolts in the floor. In the dim light of the interior, the impression created by the cargo looked for all the world like a series of stiffened corpses propped upright in rows, each held in place by a soiled collar. Packsacks and bedrolls were tied to the floor all around us. This was a new experience in travel.

At Hay River we dropped down through several layers of dense cloud and fog before breaking into the clear. The moon was almost at the full, and black spruce cut the skyline like jagged gapped teeth. The river was still gripped in winter's ice except for two sizeable areas containing patchworks of sharp-angled blocks of ice. The landing produced a peculiar sliding sensation, as though the plane's landing gear was somehow out of control. We shortly learned that the flaps had failed to come down and that the excessive impact on the high-speed landing had caused a loss of hydraulic fluid up front. Passengers were asked to huddle well back in the tail of the plane while the hydraulic facility at the front was restored to its original condition.

The airport lounge was closed; no chance for a cup of coffee. Some of us took a short brisk walk in the cold air and then returned to the plane. We took off rather sluggishly, I thought, and headed north again. A distinct broad ribbon of salmon-pink now suffused the sky to the northeast above the straight horizon. We had unconsciously passed into the early morning of May 28. The other passengers had fallen asleep in a variety of attitudes. At Norman Wells the power had gone off, the petrol pumps refused to operate and we were unable to refuel. I heard the copilot say that he was sure that we could make it anyway, so in a moment we were airborne again, heading overland for Melville Island off the Arctic coast.

It was lighter now. I looked down upon a pastel-tan background into which were sharply etched the flowing lines of drainage systems radiating like slender strands of filamentous algae urged by a gentle current. Then miles of newly fallen snow, giving the landscape a subdued pale mantle with the geographic features barely expressed. This, in turn, gave way to patterns of lakes, juxtaposed with sinuous streams and frost polygons to form intriguing relationships more fascinating than any I have seen conjured by the mind of man and set to canvas. At times, stark and bold; more often, subtle and suggestive. Always, compelling. A magnificent far-flung canvas of infinite variety.

We arrived at Marie Bay, Melville Island, in brilliant sunshine, landing on a long strip of ocean ice in the bay. This was a temporary oil exploration camp with all the attendant gear of heavy equipment, bunkhouses, cookshack, piles of lumber, windrows of red oil drums. I had a fine meal in the cookhouse and then radioed Resolute Bay, stating that I was awaiting a ride to that point which would be our nearest settlement, 350 miles to the south of Bathurst Island. A plane would be coming through from Mold Bay on Prince Patrick Island. It would pick me up the following day so I had an opportunity to look around a bit.

The camp was in the initial stages of moving to another location. It was interesting to see a large helicopter hover over a bunkhouse while some of the crew attached cables to it. Then, on signal, the bunkhouse was quickly raised vertically into the air and flown away, dangling at the end of a slender thread like a spider spinning the first almost transparent lines of its web. Drilling rigs were transported in two loads, first the base, after laying aside the superstructure, and then the superstructure itself. A large Caterpillar tractor was busily hauling in an enormous load of snow to be thawed for the fresh-water supply.

On leaving Marie Bay the next day, several herds of dark, stocky muskoxen were silhouetted sharply against the white expanse of snow on Melville, as were a number of paler, slimmer Peary caribou. The weather was socked in all the way to Resolute Bay on Cornwallis Island but cleared just enough to afford a landing there. Resolute seemed still in midwinter with enormous piles of snow everywhere. It was piled ten feet high against the sheet iron buildings which were painted a bright orange, presumably to assist aircraft in spotting the settlement in fog or snowstorms. I could make out two short streets of red or orange buildings, the post office, RCMP barracks and a USA weather office. We finally arrived at a very long, low metal building, The Muskox Inn, where visitors put up.

I registered, was given a pair of sheets and shown to a room with four steel bunks. There were no blankets. You were expected to use the sheets with your bedroll. The dining room was just down a narrow hallway. Prices? Simple; they had only one. Breakfast, $10; lunch, $10; dinner, $10; sleeping accommodation, $10. Eat all you want. I had a delicious meal with an unexpected selection in this tiny settlement, so very remote from civilization far to the south.

While enjoying the meal I heard a plane zoom over. By the sound, it was a large one, a jet. Shortly, a gentleman entered the dining room from the hall and sat down on the stool next to me.

"Would you care for lunch, sir?" the young waiter asked politely.

"I dinna think so," the stranger replied in a broad Scotch accent. "I ate on t' plane. But maybe I'll just 'ave a coffee and a wee doughnut." The waiter served him. He seemed to enjoy the coffee, then turned to the waiter and said appreciatively, while fumbling in his heavy tweed pants for his wallet, "That's fine coffee, lad, and what do I owe you?"

"Just the usual. Ten dollars."

"Hey, just a wee minute! I had only a coffee and doughnut!"

"We have just the one price, sir. It's posted on the wall there. You can eat all you wish. Would you like something else, sir?" The stranger handed the waiter a tenner and stomped down the narrow hallway.

Jan, a confident young Dutch pilot, took me out to Bathurst Island the next day in an Otter. We were socked in most of the way and I couldn't see a thing clearly.

"The tiny camp is very hard to see in this stuff," advised Jan. "I'm not sure that we'll be able to spot it. Just a minute! There it is!"

I followed his pointing finger and could just make out a small square-looking red patch in a sea of white. Then, as Jan circled for a landing, I saw a second greyish-white object some distance from the red one. We came down between the two and bumped to a halt. I jumped out and shook hands vigorously with Stu, standing between two others on the rude improvised landing strip of packed

The Bathurst Party. Right to left: David Gray, Phil Taylor, Dr. Dave Parmelee, S.D. MacDonald, Dr. Cy Hampson.

snow. The leader of the expedition was tall, black-bearded and grinning broadly in the warmth of his welcome.

"Glad you made it, Cy! Welcome to the last of the unexplored High Arctic islands! Hi, Jan. Good trip?"

"I thought we would have to turn back for a bit there, but spotted you at the last minute."

"Man, what eyes your pilot has," I offered. "I couldn't see a thing."

"You know Phil, Cy. And I think that you've met Dave, Dave Parmelee. Right?" Stu said, turning to his companions.

"Yes, we met in Edmonton, once. Hi, Phil. Hi, Dave. Good to see you both." We shook hands warmly. "Say, what about Dave Gray? And how do you keep the two Daves straight?"

"Nothing to it," laughed Phil. "Dave Parmelee is Dave and Dave Gray is David. David's out with his muskoxen right now."

Phil Taylor was a compact, energetic, outgoing young man with rounded features and a wealth of enthusiasm. He would be working with MacDonald as an assistant. Parmelee was sharper featured, a mature, experienced northern researcher who had dedicated much of his life to unravelling the mysteries of arctic-breeding shorebirds.

"Coffee, Jan?" Stu invited.

"You bet."

I threw my gear out of the plane and Phil gave me a hand taking it into the red hut, which proved to be a double-walled Quonset affair made of nylon and equipped with an oil-burning heater and six single bunk beds.

"Just throw your bedroll on that empty bunk and we'll go over and have coffee."

The white structure was the double-walled cooktent. It was complete with table, three-burner gasoline stove, several cases of supplies and packing boxes for

Base camp on Bathurst Island, 1968. Bunkhouse, left centre; cooktent, far right.

chairs. Plastic windows in both layers of nylon, together with the white colouration of the tent, admitted plenty of light. It proved very snug once the stove was operating.

"We've sort of split the stuff in the interests of safety," Stu explained. "If we had a fire and everything was in one place, it could be catastrophic."

"You think of everything," I said. "No point in having all of your eggs in one basket. What are you going to work on this trip, Dave? Isn't this about your seventh expedition to the Canadian Arctic?"

"Seven it is. I hope to work on the first Canadian study of the little sanderling. I hope they turn up in numbers. Stu and Phil are going to work on rock ptarmigan. Boy, you should have been here yesterday."

"What happened?"

"We watched a wolf kill a big bull muskox. Saw the whole thing, believe it or not."

"I'll believe it, coming from you. Wolf or wolves, Dave?"

"One wolf. Single-handed." Dave's voice held suppressed excitement.

"Single-handed, you say. How did he go about it? Hamstring him first?"

"No way! Old Bloodface, that's what we call him, hasn't read any of the standard works on wolf predation. He didn't stalk the bull, he didn't hamstring him, nor did he sever the jugulars. He didn't do anything that wolves are supposed to do when killing prey animals."

"What did he do then? I'm sitting on the edge of this box!"

"It's one for the books, this one. The muskox had been over there across the valley, alone for two or three days. We were checking him one morning after breakfast with our field glasses when we saw the wolf trot over the hill a little to

Dave Parmelee's birthday. The shorebird expert.

the left of the muskox. He spotted the bull and trotted right up to him. The two looked at each other for a second or two and then the wolf darted forward as if to say, 'Well, it's either you or I, so let's go!'

"The muskox backed off a few steps, lowering his head and hooking sideways with his horns. The wolf stopped and the muskox charged. Bloodface danced quickly backwards ahead of the horns and then whipped around behind the bull. The bull turned and the manoeuvres were repeated."

"Whew! What a sight that must have been!"

Dave continued. "This seesaw battle went on for quite some time, but after awhile it looked as though the muskox was slowing down a bit. The wolf seemed as fresh as ever. Then we noticed that the wolf was manoeuvring the bull. Whenever the muskox charged, he was always working uphill. With Bloodface in front of him, of course. The wolf's strategy was clearly to take advantage of the terrain, forcing the heavy ungulate to charge against the grade while he made his moves downhill. Old Bloodface was in complete charge all the way. No doubt about that."

"So he finally tired the bull out?"

"Partly that, but the important thing was that he slowed him down. Then we saw Bloodface take a great leap at the bull's head. We couldn't see what happened because the wolf was on the far side, but the bull began to shake his head from side to side. Bloodface danced around to the same side and leaped in again. This time he locked his jaws on the bull's throat and just hung on. The muskox tried to hook and shake him off but he wouldn't let go. Finally, the muskox sank to his knees, still struggling, but it was all over. In a few minutes, the bull lay still."

"What an experience! And new insight into wolf behaviour, for sure."

But Dave hadn't finished. "We felt that something must have happened to make the bull shake his head so vigorously after the wolf had leaped in the first time, so we jumped onto the snowmobile and toboggan and went over to have a look. When we got there, Bloodface had opened the chest cavity and begun to feed.

"When we cut the motor close to the carcass, Bloodface left it, walked away a dozen steps and sat down, watching us. We checked the carcass and found that he had taken out one of the bull's eyes, undoubtedly in that great leap. Then he had attacked from the blind side. And looking at that bull's head with those needle-sharp horns curved inward towards the eyes, I still can't figure how he did it!"

"Then we loaded the dead bull onto the toboggan, tied it down and started back to camp," Phil Taylor cut in, his dark eyes flashing in his round face. "And what do you think happened?" he challenged.

"You had stolen his kill, so he moved in on you and you had to shoot him?"

"Nothing of the kind! He followed us all the way back, dancing around us in circles at times. He even jumped in and grabbed a rope that was trailing behind the toboggan. The bull's still out behind our sleeping tent. We hope to get some good pictures," Phil concluded.

"You're right. I should have been here yesterday."

"Any idea what you'd like to work on, Cy?" Dave asked.

"Sure thing. Muskoxen kills by wolves! No, I don't know what's here. We were socked in nearly the whole way from Resolute. I liked the look of the caribou on Melville and I saw a couple of Arctic hares there, too. Any of them here?"

"Quite a few of both," Dave replied.

"Then I'd like to do something on their locomotion if I can. And give any of you fellows a hand if you need one."

"Thanks, Cy," Stu said. "We'll probably be calling on you, especially at the beginning when we're trying to locate breeding territories and the like. You will enjoy the hares. No doubt about that."

We finished a second cup of coffee along with crackers and peanut butter. Mine had grown cold listening to the story of Bloodface. Jan got up to go and we shook hands all round. The Otter lifted off after a very short run and we all repaired to the red Parkall bunktent. I unpacked my stuff, put sheets in my bedroll and got out the portable typewriter to bring my notes up to date. It was nearly time to turn in.

During the night, I was awakened several times by the billowing, cracking and popping of the fabric on our bunktent, but when I got up the wind had subsided somewhat. Phil was already up gathering the basic data on temperature, wind direction, wind velocity and the like. The other bunks were empty as well, and I thought I heard the clatter of a pan in the cooktent. I dressed hastily and stepped outside to have my first good look at the camp and its setting. I recalled that according to the large-scale map, Bathurst Island is located astride the 75th

Rock ptarmigan are the most northerly of the ptarmigan, occurring as far north as the limit of land.

parallel, roughly fifteen degrees from the North Pole. It is approximately 140 miles long and seventy miles in width, on the average, with a narrower section south of centre bounded by Bracebridge Inlet on one side and Goodsir Inlet on the other. Our headquarters, according to information received, was to be about halfway between the two inlets.

The camp was located roughly on the east-facing brow of a rounded hill with a good view in all directions. To the south lay a very broad, low valley, extending to the horizon on the left and right. This would be the Polar Bear Pass marked on the map that Stu had sent me. Another, but much closer valley lay to the east with rounded hills perhaps two or three miles distant, and on the far side the ground rose to the north in a series of hills, with what appeared to be a small drainage basin issuing from them. The land sloped gently to the west where, about a hundred yards away, stood a group of snowed-in oil drums with a sort of blue and white flag fluttering in the very cold breeze. While there were great expanses of snow in all directions, here and there, on the crowns of hills or steep slopes, bits of dark tundra showed. There was certainly little evidence of spring.

I hurried across to the cooktent where everybody had already gathered.

"Hi, Cy!" Stu hailed. "I don't think that you've met David Gray. Remember, he's David, not Dave."

"Right. Good to know you, David. You're working on muskoxen, I think Phil said."

"Yes, I am." David spoke quietly, confidently. He was a tall slim young man with a great shock of dark hair and lean face with squarish chin. "I'm having the time of my life. This is a great place to be." His eyes were glowing. He meant it.

"You were out with your muskoxen yesterday?"

"Yes, I plan to spend every day with them. I'm particularly interested in their behaviour under all sorts of conditions. The fellows here have agreed not to disturb my herds any more than necessary when going about their own work." He spoke directly.

"Good idea. You'll have to let me know where your herds are located and I'll try to avoid them. Pretty cold out there in the driving wind right now, isn't it?"

"Not bad at all. Well, I've got to be off." And he proceeded to put together some lunch, presumably for the day.

"Say, where's the biffy, Dave?"

"Haven't you seen the flag? The biffy's right under it. The snuggest, cosiest biffy within fifteen degrees of the North Pole! Guaranteed to discourage meditation and accelerate defecation. You haven't lived, Cy, until you've tried 'er out in the wind!" Dave was smiling broadly, his eyes twinkling.

I tried 'er out in the wind. Right after a welcome breakfast of delicious porridge, pancakes with syrup, and two cups of scalding coffee. The biffy was composed of a forty-five-gallon oil drum, open at the top with a half-sheet of plywood over it. A suitable hole had been carved in the plywood and the other half-sheet erected on the north side, the windward side. The newly blown snow was easily brushed aside.

Seated there in the open, with a biting wind off the pole and the temperature well below freezing, I didn't feel disposed to meditate.

The Arctic Hare

Arctic explorers left many fascinating accounts of the great white hare which they encountered in so many remote regions of the High Arctic. At times, the herds of these hardy animals were so large that as they travelled from place to place they left behind hard-packed trails in the snow a dozen feet wide. One can only guess at their incredible numbers at such times, and when one examines the sparse growth of vegetation at such high latitudes, it does not seem possible that the land could support them. The Arctic hare must be extremely efficient in its utilization of the little that is available, especially during the very long, cold winters when all growth is suspended.

The explorer Bay, while examining the remote west coast of Ellesmere Island back at the turn of the century, writes of them: " . . . the place was teeming with them. They ran about as though they had taken leave of their senses. It was in the midst of the pairing season, and we supposed that they had lost their heads from love. It is a thing which happens to others besides hares."

A longer account of these amazing herds of hares on the same island was given by Sverdrup about two years later: "I now put on my ski, and set off as hard as I could for camp. Round about me, both before and after I crossed the narrow canyon, the slopes were swarming with Hares which were out foraging.

"I drove on next day at our usual time; again passed the herd at about the same place I had seen it before, and [we] were soon down through the narrow canyon, although now and again we had to put ourselves, all four, to the sledges to bring them across the stony ground.

"A little farther on, we drove past a Polar herd on the other side of the valley. Some of the animals were standing on the precipice; others were climbing up and down the steep stony slopes, looking like flies on a wall. When we were right under them, they formed up on the precipice, and stood glaring down at us.

" . . . All these Hare tracks were one of the most extraordinary things I had ever seen; never could I have imagined that their pads would be capable of making such enormous runs.

"The farther we went, the more numerous were the Hares; and later in the afternoon, when we reached the lower valley, they seemed to be conjured forth from the slopes as if by magic. There were such legions of them and they scurried about so in all directions along the valley, and backwards and forwards across it, the Dogs became absolutely unmanageable. It was impossible to keep them in check; they gave chase time after time; and the Hares themselves were so dazed that they had not the wit to keep out of the way. They did not appear to be

Arctic hare resting between feeding intervals.

afraid; they hopped about only a few yards in front of the teams. At last, after the dogs had bolted after them time and time again, we finally landed in the steep bank of a river . . . .

"As if to incite the Dogs to the utmost, the Hares came out and settled down a few yards from them, and then stood on two legs and stared at us. The Dogs yelped and snapped and made a shocking clamour . . . and the Hares? They never moved from the spot! What were we to do with them?"

During the First World War, Freuchen was the only white man living in Thule, Greenland. At this time, he and his Eskimo friends made many long trips across Smith Sound to Ellesmere Island to secure large numbers of hares in order to make stockings out of their soft fur. Before the first journey, the hunters told Freuchen that in Ellesmere, "there are so many hares that it looks as if the ground has lice!" The party was not disappointed; hares were plentiful, indeed.

A third of a century later, the arctic explorer Ross recorded yet another use of the Greenland hare by the natives: "Amongst the inhabitants of Greenland, one Esquimaux woman was found who spun some of the beautiful white wool of the hare into a thread, and knitted several pairs of gloves; one pair of which . . . came into my possession beautifully white. It very much resembles the Angola [Angora] wool, but is still more soft."

I first saw Arctic hares on Bathurst Island on June 5 when hiking up a steep, partially snow-clad slope northeast of our camp. While the snow in the extensive low-lying areas and on the lee side of hills was still very deep, the crowns of some of the rounded hills and windward slopes were partly snow-free. The strong wind out of the north was biting. I had paused, breathing hard, to admire the enormous snowdrifts and snow cornices in the dry stream bed to my right. On the far side of the valley a series of vertical spaced grey columns of rock reared into the heavens like strange, silent monoliths.

About to step forward again, I noticed a white object on the tundra just ahead. It was an Arctic hare, perfectly motionless. He had his buttocks tucked in securely against the slope and his body flattened on the gravel, with forelimbs

Hare digging for roots along the margins of melting snow where the soil was moist.

extended. The ears lay down, almost flattened against the back. The head was drawn back against the shoulders and the eyes were closed, except for a narrow dark slit. When I crept to within four or five feet of him, he slowly rose into a sitting position and yawned prodigiously, exposing the long, protruding upper pair of incisors. The hare was clothed completely in very long, white, silky fur except for the tips of the ears, which were black.

On looking carefully at the area around the hare, I saw nothing but bare patches of sand and gravel interspersed with slowly retreating snowdrifts. What could the hare possibly find to eat in such surroundings?

As though reading my thoughts, the hare suddenly raised its ears, extended its front legs as far as possible and then moved its body forward in a prodigious stretch, with the hind toes still in contact with the ground. I could feel the tension in its rigid body. It then hopped leisurely to the edge of a snowdrift and proceeded to extend its front legs as far forward as possible along the ground. Then it drew the front feet back alternately, with claws dragging through the moistened soil. To my great surprise, a number of slender strands of black roots clung to its claws. The hare methodically transferred the roots to its mouth, chewed them thoroughly and then swallowed them. Later, throughout the early summer before there was any significant growth in the local plants, the hares gathered the roots of Arctic willow, purple saxifrage, dryas, sorrel, ground-hugging cinquefoil and Arctic poppy in much the same way.

A still different method of food gathering was observed on the same day. The hare had hopped up onto the edge of a very thin patch of crusted snow. The extended front limbs were used to beat a tattoo upon the surface of the crust, shattering it into small fragments. These were then removed by use of the forepaws and mouth. Then the hare pushed his nose as deeply into the soil as

Hare taking soft faeces from anus in order to put them through the digestive tract a second time.

possible and nipped off roots, which he brought to the surface and chewed audibly. The importance of the slender protruding incisors in this process was apparent.

On one occasion, after resting for some time, the hare roused, yawned, stretched and then raised his body almost vertically over the front legs, with the rump still on the ground. The front paws were spread and the hare reached down between them to his anal region. About two inches of the end of the digestive tract was extruded while the hare ate the soft faeces as they emerged from the anus. As with most hares and rabbits, food is put through the digestive system twice in order to extract the maximum amount of nutrients from it. The first faeces are soft; the final ones, small compact marbles. The initial soft faeces contain the important food elements that result from the breakdown of cellulose by the bacteria of the caecum. There is little waste of essential elements. This practice is so important to these animals that they tend to die in two or three weeks if prevented from reaching the anus.

The fact that the hare makes use of so many different kinds of roots and rootlets is undoubtedly important in enabling him to survive in the event that some of them are in short supply.

On the other hand, many muskoxen had failed to survive. During the many hundreds of miles of tundra that we covered in our search for birds and mammals, we ran across many carcasses of muskoxen, several of them intact. We turned up no yearling muskoxen and no calves that year, which indicated that the

*Hare grooming itself
and ingesting some
of the fur,
presumably for its
protein content.*

muskoxen had undergone a period in which reproduction did not occur. The low-lying areas were heavily blanketed in deep snow. It turned out that though there was life-preserving vegetation beneath the snow, the muskoxen were not able to paw down through the depths to obtain it. It appeared that many of them had died of starvation. It seemed apparent that the island had experienced at least two consecutive winter seasons that had proven very adverse for the muskoxen. It is well known that many forms of wildlife, from deer to great horned owls, fail to breed when very poorly nourished.

The exposed areas of the range were in very poor condition with respect to vegetative cover. The Arctic willow had been cut back severely by foraging animals, and other plants had not yet begun to produce new leaves and stems. The rounded hills, with their cover of sand and gravel, were sparsely dotted with plants at best and, since the area was in fact a northern desert, it could produce but limited amounts of vegetation due to the porous nature of the sand and gravel cover, the sloping contours and the very limited precipitation. It seemed ironic that the low-lying plains and valleys, which held the moisture longer and produced quantities of forage, were so deeply covered with snow that the plant material was only available to starving herds for a very short period of the year. The hares survived and later produced young since they were able to utilize efficiently what food there was available in the form of roots.

The Arctic hares of Bathurst Island had still another behaviour pattern that protected them against the very severe winters of such far northern latitudes. On June 6, I found two snow burrows excavated by them. Both were dug deep into snowdrifts on a southeast-facing slope, about thirty feet below the summit of a broad hill. A good deal of hare fur clung to the roofs of the tunnels and there were many droppings on the floor, both suggesting extensive use during the winter. The burrows ran in north and northwesterly directions from the entrances, which faced downwind from the strong prevailing nor'westers.

Excavation of the first revealed that it ran back just over six feet to a rounded chamber showing many claw marks from shaping it. The other was a little over five feet in length.

During our stay on the island, we witnessed very few interactions between the Arctic hare and other animals present. While wolves passed through our study area on numerous occasions, we did not record a single instance in which they pursued hares, even though they often trotted by within a few feet of them. On one occasion, Parmelee observed an Arctic fox stalking two hares. The hares were feeding on the side of a ridge as the fox approached. Its eyes fixed on the hares, the fox moved slowly and cautiously forward, pausing motionless at intervals. The hares continued feeding unconcernedly. The fox finally reached a point thirty or forty yards away and then made a sudden dash forward. The hares spotted him the moment he broke and fled to the top of the ridge, using powerful kangaroo-like hops on their hind legs. Then, having gained the summit, the hares stood bolt upright for several seconds, outlined against the sky. The fox, easily outdistanced, turned away and trotted down the near slope.

But on June 15, MacDonald, Taylor and I enjoyed a ringside seat in what appeared to be a duel between a snowy owl and Arctic hare on a south-facing slope along the dry riverbed. We first noticed the owl as it flashed down upon the hare, missing it by inches as the hare dodged to one side at the last split-second. The owl landed in the snow, scarcely a yard away. After the hare had hopped some distance away, the owl again took wing in pursuit. Again the hare dodged nimbly to one side and the owl drove straight ahead, ploughing into the snow in a white cloud of crystals, a dozen yards from her target. To our utter amazement, the hare then turned and hopped to within a yard of the would-be predator. Could the hare have spoken the language one felt that she might have been saying, "Now, is that really the best you can do?"

At this juncture, the owl changed her strategy and began to pursue the hare on foot. But she was very awkward, bouncing over the snow in short, heavy, ungainly jumps. The hare evaded her easily, unconcernedly. At length, the owl gave up this tactic and flew to the top of a snowbank nearby where she perched, eyeing the hare. The owl remained motionless until the hare had considerably increased the intervening distance by hopping towards an exposed patch on the slope from which the wind had removed much of the snow. The owl took to the air again but narrowly missed as before. Seeming to prefer to attack from a greater distance where the momentum of her diving strategy was increased, the owl repeated the manoeuvre again and again with the same results. Then the hare hopped round and round the base of a large snowdrift with the owl perched on the summit, its head swivelling to follow the hare. Then, still another stoop, only to close her talons on loose insubstantial crystals of frozen snow.

It was really a deadly serious game. One miscalculation on the part of the hare and the owl would dine. While the hare appeared to toy with his powerful adversary, he was nevertheless very much on the alert and apparently aware of the bird's intentions. What seemed to leave the owl completely bewildered as the game proceeded was the manner in which the intended victim approached her at

At age six weeks, the leveret is assuming the white garb which it will thereafter retain.

the conclusion of an unsuccessful attempt. We watched the drama for the better part of an hour at the end of which the owl flew off over the tundra, empty-taloned.

After the owl had departed, the hare lowered the back of the head, neck and shoulders into the snow, the spine twisted and both hind feet still in contact with the snow. At this stage, the hare reminded me of a coyote beginning to roll upon a shrew or bit of old rotting meat. But the hare let go with the hind feet and was immediately rolling in the snow in the manner of a horse, with all four feet flailing the air vigorously. Head, neck and shoulders were moved energetically from side to side as they rubbed against the snow.

June 28 was another red-letter day. David Gray came running to the bunktent with news.

"Hey fellows, there is a hare out behind the biffy with four youngsters!" David had spotted it while cogitating.

We all rushed out. There was no white hare in sight but David pointed to the area behind the biffy where he had seen them and then left to continue his work on his muskoxen study. It was a long painstaking job finding them, as they had completely "frozen" into a bare area of tundra. Two hours of searching, foot by foot, turned up the first one. It was crouched in a very shallow depression with rump to the sun. I turned my eyes away for two or three minutes while digging my camera out of my pack and had to look very hard again to spot the leveret, even though it hadn't moved. Had its head been facing the low sun, I might have caught a glint in the eye but it was facing in the opposite direction. I was to learn that this was a characteristic attitude of a hiding leveret.

The thoroughly delightful little fellow was barely six inches long, generally sandy-buff in colouration and undoubtedly not more than a few days old at most. The soft fur in the regions of the shoulders and forehead was still slightly wavy

Snowy owl in full flight.

and curly. Dark brown guard hairs were numerous, silky and tipped with buff. The grizzled ears, lying against the back of the neck, were black-tipped with narrow creamy margins, while there were light buffy areas around the eyes. The forehead appeared relatively broad; the nose, short and blunt.

The leveret did not move a muscle when I photographed him from a foot away. I found a second member of the litter an hour later, barely two dozen strides away from the first. In attitude, colouration and behaviour he might have been a carbon copy of the first. Not wishing to prevent the mother from attending to the needs of her young family, I did not attempt to find the others. As I was to learn, there was little need for concern in this regard. She would be absent for a very considerable time.

The mother hare did not return to nurse the young leverets for several hours and then she stayed but a very few minutes before leaving. Unwilling to disturb her in the care of her youngsters, I watched her from a distance with field glasses but could make out little detail. As she hopped to the area where she had left her litter, her size and white colouration made her stand out on the tundra, much of which was now brown and snow-free. It suddenly struck me that we had not seen the white male for two or three weeks. And when the mother left the nursing area, she left alone. The leverets did not follow her as the rock ptarmigan chicks

*Snowy owl
nearing target.*

followed the hen. In addition, the hare did not pause to feed until she was at least a quarter of a mile away.

I tried to put the pieces together. The young brownish leverets were perfectly camouflaged to blend in with the bare area of tundra where she had left them. What I took to be the conspicuous white male had departed and the equally conspicuous female remained with her litter only long enough to nurse them briefly. Nor did she forage in their vicinity. The colouration and behaviour of the leverets, left by themselves, would make it very difficult for predators like owls, foxes, glaucous gulls and jaegers to spot them. Their only defence at this stage lay in their virtual invisibility. On the other hand, the presence of the conspicuous adults could serve to draw the attention of predators to the vulnerable youngsters.

It was not difficult to conjure up the plight of very young hares following clearly visible white objects over the tundra. White objects which, while attracting the attention of predators to the leverets, were incapable of defending them against attack. The adults with their speed and manoeuvrability were able to take care of themselves and could afford to remain white; the helpless leverets were in need of effective camouflage.

The following evening proved an exciting one. I left camp loaded down with several cameras and a heavy tripod in the hope of recording some sequences on the attractive leverets. I made a careful search of the area where they had last been seen and finally found one of the youngsters crouched in a small depression in a patch of purple saxifrage. As usual, his eyes were turned away from the sun. I nudged him gently with my forefinger, at which he obligingly turned about, but before I could bring him into focus with my camera, he had done another switch and lay in the original position again. A half-dozen repetitions of the manoeuvre convinced me that he had his own predilections about the proper way in which leverets ought to arrange themselves when resting upon the open tundra. We mere humans had much to learn.

The mother was not in the vicinity. I began skirting the surrounding area thoroughly in search of her as I dearly wanted photographs of her with the leverets. Using field glasses, a careful check of all the hills, ridges and valleys within a mile of the youngster failed to turn her up. I waited for her to show, as this was about the time she had nursed her offspring on the preceding day.

An hour or so later, I noticed a small white spot on the steep slope of the riverbank on the far side of the valley to the northeast. Field glasses revealed that the small spot was indeed an Arctic hare, and since there were no others anywhere to be seen, I decided that there was a very remote chance that this was the mother of the litter.

The odds seemed exceptionally low in light of the fact that the hitherto dry riverbed had become a raging torrent a few days earlier when the temporary river had "gone out" from the accumulation of meltwater in the surrounding hills. Four muskoxen had attempted to cross it two days earlier. They had succeeded in reaching a point near the middle of the swift stream but then turned back, defeated, and waded heavily to shore. The mother hare, if this were in fact she, was feeding on the other side of the river, currently in spate. I sat down to await developments, should any occur.

The hare continued feeding and resting on the far side of the river for another three hours, then suddenly started down the slope in the direction of a point where I had several times seen hares cross the valley before the river had begun to run. As she approached the stream, she paused several times with head held high as though surveying the terrain. At length, she reached the first of the three branches of the stream and attempted to cross at a broad point, hopping into the water until it was well up on her sides. Apparently unhappy with this situation, she returned to her original point of departure and hopped upstream to a point where the river ran over a gravel bar, forming swift but shallow rapids. These she crossed in long leaps without difficulty. The major stream, resulting from the confluence of two substantial forks, lay directly ahead. What would she do?

She paused in the gravel a few yards from the edge of the stream for a quarter of an hour, squatting with feet tucked under her, back arched and ears flat in the typical attitude assumed by hares when resting or sleeping. She then aroused, rose to a sitting position on her haunches, looked around circumspectly, I thought, and hopped to the edge of the river. With a great leap, she plunged in. Though the racing water was deep, she continued leaping forward until she had reached midstream. Then, like the muskoxen earlier, she turned back, reaching the bank several yards below her point of departure. The current was very strong.

At this point, the hare stood peering up and down the stream for a minute or two before hopping upstream to a point where an expanse of ice extended well into the water from the shore. She started across here but the thin ice gave way and she was forced to retreat to the bank again. She then stretched her neck upward, looking up and down the river for a few seconds before making a wide detour upstream to a spot beyond the intersection of the two major tributaries. Here, though the streams were narrower, they appeared quite deep and swift.

The Arctic hare finally nursed her leverets.

Without hesitation, the hare plunged in and continued leaping forward, the water over her shoulders at times and carrying her rapidly downstream. But she finally reached the gravel bar between the two streams and leaped out. She spent the next few minutes dressing down her wet fur and resting. Then into the second stream, scrambling and swimming to the near bank. Her belly fur was thoroughly soaked and dripping on arrival.

The hare immediately set off in a westerly direction up the steep bank. She shortly appeared on the upper slope of the hill west of camp, where she paused to groom herself and feed for a few minutes. She then hopped directly towards the shallow depression where the leverets had first been seen.

Uttering a low growling sound, she began running in a small circle perhaps ten or twelve feet in diameter. On the instant, small brown bodies began leaping towards her from different directions. In seconds, she was being pursued closely by four youngsters. She then stopped abruptly, settling on her haunches with ears back and hind legs spread widely. As the youngsters crowded beneath her, reaching up to nurse, she moved her front legs farther apart to accommodate them. The leverets suckled for approximately four minutes before the mother hopped forward. Three of the youngsters immediately scattered and settled in shallow depressions, melting into the tundra. The fourth followed her for only a few strides and then the mother lengthened her stride, at length disappearing over the brow of a hill to the west.

I did not see the mother with the leverets on the following two days, but did observe two of them feeding upon willow shoots and the blossoms of purple saxifrage. With the passage of time, the leverets grew steadily paler, apparently by a combination of the bleaching of the original fur and the growth of long white hairs from below. Within five and a half weeks, they were almost white except for areas on the head and ears and a pale greyish area back of the shoulders. When winter returned on August 12, these leverets were already white. The adults remained white throughout the season except for the very small rusty-brown areas about the head, noted in late June and July.

The seasonal colouration of the hares was intriguing. In these far northern latitudes in which the summers are so very short, the adult hares remain white throughout the year. The brown young are born in the short summer when most of the tundra is also brown, but they turn steadily lighter in colour as they mature and as winter approaches. By the time the snows begin to blanket the tundra, they are white and will never again be brown.

In these high latitudes, during the very brief period when the young hares are most vulnerable to predators, they are the colour they ought to be if they are to survive. The same species, living on the mainland farther south, experience seasonal colour changes of varying degrees.

The Arctic hare is a cursorial mammal, a speedster well able to outdistance its terrestrial predators, the wolf and fox. The long, powerfully muscled hind limbs are able to propel it forward at great speed. The flexibility of the spine allows the same rear limbs to pass well forward of the shorter front ones when running, thus greatly increasing the length of stride which, in turn, increases velocity. The efficiency of this gait is again increased by lining up the two front feet, one directly behind the other, so that the hind limbs can pass around the forelimbs without the necessity of expending additional energy in moving the former outward and then inward in order to accomplish the same passage.

Again, when at higher and higher speeds, the hare's tracks clearly showed that all four limbs tended to function more and more closely to a midline with the four feet striking the snow one after another. This pattern reduces the braking effect resulting from the hind feet coming down in unison, side by side, as the hare generally practised when moving slowly. In addition, higher speeds were always accompanied by longer and longer forward leaps.

From protruding canine teeth to efficient running form to effective colouration, general alertness, behaviourial patterns and efficient use of the very limited resources of the tundra, the Arctic hare is superbly equipped to deal with the exigencies imposed by the exacting environment in which it lives.

The Elusive Sanderling

I hurried over to the cooktent early one very cold morning shortly after arriving on Bathurst, only to find that Dave Parmelee was already there and finishing his second cup of coffee. The wind out of the north was billowing the walls of the tent and Parmelee had his fingers cupped around his coffee mug as though to warm them.

"Cold morning, Cy," he greeted me. "There's hot coffee in the pot." I poured myself a full mug and added a squirt or two of evaporated milk from the perforated can on our rude table. The aroma of the steaming liquid was inviting.

"And what's up for today, Dave?"

"Combing the hills back there in the hope of finding some displaying sanderlings," he said, waving a hand in the direction of the hills to the north. "Sure hope some of them breed here this year." His voice was eager, hopeful.

"Dave, this is your seventh season in the Arctic?"

"Yes, it is. Could hardly finish the term down in Kansas. I had to get back home."

"And you're still working on shorebirds?"

"Yes siree!" There was no doubting his enthusiasm.

"Why shorebirds, Dave?"

"I'm not sure. Stu's a grouse man, as you know. And I know a fellow who works on nothing but raptores. He once told me that he figured that the birds of prey have had a pretty rough deal over the years with so many people shooting them on sight. He feels that they have a pretty important place in the web of life and he wants to get that across to people in the hope that the raptores will some day receive more consideration and protection."

"But that's not the way you feel about shorebirds?"

"No, but they're exciting. Look at the long migrations of some of them. From the High Arctic here to South America, to Australia, to New Zealand, to Africa. Man!" He topped up his coffee mug from the pot on the burner, his eyes shining, and continued. "Maybe I'm a bit envious of their easy movement over the globe."

"But you specialize in arctic breeders?"

"You bet. We know so little about many of them. How do they fit in here? What do we know about their behaviour patterns on their remote breeding grounds? Everyone of them is a challenge."

"But why sanderlings, now, Dave?"

"Simply because so many bits of conflicting information about them has been coming forth, that's all. Ever since Manniche's study of them in Greenland back about the turn of the century. We just have to sort out as much of the truth as we can. Did you know that there has been no previous study of these birds in Canada? This will be the first." And Dave hesitated, looking inwardly, visibly hoping against hope. Then he finished.

"That is, if we find enough birds to do a proper study."

I realized that Parmelee's projected study was a long shot, but I knew that he would tackle this self-assigned study with his customary enthusiasm, energy and thoroughness.

The sanderling is a rather chunky, short-legged sandpiper. In breeding plumage it is readily recognized by its black bill and legs, light underparts, rusty head, neck and upper breast and the prominent white stripe in the wing, conspicuous in flight. However, most people identify these widely distributed sandpipers during migration and in winter by their very pale colouration contrasting with that of most other shorebirds. They may be seen at this time along many of our sandy or muddy beaches, running swiftly before advancing waves as though on well-greased wheels. When the waves reach their limit and turn, the sanderlings quickly reverse direction and pursue them as they ebb. The sanderling breeds on arctic tundra around the world, but winters in the southern half of North America, Chile, Argentina, the West Indies, South Africa, Australia and New Zealand. The thought of a few ounces of flesh and feathers flying those astounding distances is staggering.

By far the greatest item of controversy regarding the reproductive behaviour of the sanderling concerned the role of the sexes. The Greenland study concluded that only the female incubates the eggs and only she takes charge of the new broods. While other observers supported this view, still others came down on the opposite side. "Male sanderlings do incubate and they do take care of broods!"

The problem was further complicated by the fact that male and female sanderlings cannot always be readily distinguished in the field. While the female tends to be paler, there are paler males and somewhat darker females. Of course, while attempting to solve such problems, a good deal of further information tends to accumulate. That is exactly what happened in Dave's study.

I saw my first Bathurst Island sanderlings briefly on June 3 when a small flock of eight of them landed on the tundra just north of our camp. They stayed only long enough for me to determine that four of them seemed quite pale and the other four considerably darker. I guessed that the flock probably consisted of four pairs, realizing, though, that the paler members could have been males that had not yet assumed full breeding plumage.

I had spotted two curlew sandpipers on June 5 and on the 7th set out to look for them again. They had been over on a promontory that we had dubbed Hare Hill in recognition of the fact that the area seemed favoured by several Arctic hares. It was late evening with scarcely a breath of wind. Then a great fog rolled in from the southeast, creating an eerie fogbow. Land forms immediately around glowed with a ghostly light. Snow cornices along the stream bank, wind-formed patterns in the snow at my feet, and eroded surfaces were breathtaking in

Sanderling at nest. Parmelee had elected to carry out the first Canadian study of this incredible shorebird.

design and texture. In a moment the camp and its hill were completely obliterated. It was strangely exhilarating to be retracing one's tracks, totally cut off from civilization in the midst of a great silence and fathomless distances.

On June 7 I saw no fewer than fourteen sanderlings together with my first ground and flight displays of the species. They were exciting. The male, on a patch of ground partially cleared of snow by the wind, raised his scapulars, drew his head into his shoulders, lowered his tail and primary wing feathers, held his beak horizontally and pursued the female on the ground, dodging erratically about. The flight display was even more spectacular, as is the case with so many shorebirds. The male rose into the air to hover on rapidly fluttering wings for periods of two minutes or more. The beak was pointed slightly downwards and the scapulars raised. Hovering periods were interspersed with brief periods in which the extended wings were held perfectly still. The bursts of song which floated down on the freezing air were somewhat frog-like and not altogether unlike the flight song of the semipalmated sandpiper, though the latter is more twitter-like.

The flight display occurred on this day over a vast sloping snowfield in very cold temperatures with a biting breeze out of the north. Females were not in evidence, although there were altercations between males in which the defence of territory seemed to be clearly involved. I felt that a resident male was likely staking out a territory to be used subsequently for nesting purposes. This was not the case. In fact, the sanderlings forced me to rethink all of my preconceived ideas about the meaning of territory in birds.

While the party recorded many sightings of sanderlings over the next few days, their local numbers on the tundra did not appear to increase. It seemed logical to conclude that the majority of birds were merely passing through to other breeding areas. The weather was severe with strong cold winds out of the north most of the time. The snow disappeared slowly. Would the season be sufficiently long and favourable to allow small shorebirds to reproduce here at all?

On June 20 Dave Parmelee and I were working a low-lying area across the bed of a stream northeast of camp when we spotted a pale sanderling running about on a bare patch of ground as though agitated. In a moment, she took wing and flew off but returned again, appearing and sounding as excited as before. We watched her for some time, hoping that she would go to a nest, but she failed to do so. On returning the following day, we found two sanderlings in the same place. They were both running about among the sparse clumps of brown saxifrage stems and dryas as before. One appeared to be a very pale female; the other, a brightly coloured male. On several occasions the female dropped into a particular small rounded cavity which had been hollowed out in the sand and gravel, but she did not remain long. We looked around very carefully and turned up several other similar scrapes. The area looked most promising.

"This area will bear checking out," suggested Parmelee.

When we returned again the next day, a bright male and a pale female were still there, running here and there over the gravel. The male had his head tucked in against his shoulders and the feathers on his shoulders partly erected. This had to be the nesting territory. We sat down at some distance, zipped up our jackets tightly and watched the birds closely with our field glasses.

The female finally made her way to the same scrape that she had occupied on the previous day and slipped into it, wriggling and turning. She finally settled with breast low and tail elevated, displaying her white underparts. The male approached from the rear, and reaching under her, proceeded to pick out bits of nest lining, which he tossed aside. Then, he pushed his beak under her belly, raised her off the nest and both sidled off side by side for about twenty feet, rubbing shoulders at intervals. The male then stepped ahead of the female and reversed his direction, pressing his beak against the female's breast. The female's forward progress was arrested, the male mounted her and mating took place, accompanied by much wing flapping. After perhaps a minute, the female ran around in tight circles, dislodging her mate. She fluffed her feathers vigorously and the two flew off. We were so certain that this was the nest that we recorded it as Nest 1 and erected a blind about sixty yards away. With the blind in place, we started back to camp. Parmelee was elated.

"In all of your expeditions to the Arctic, Dave, have you ever before seen anything quite like that?" I asked.

"No, I haven't," he replied with a broad smile. "That was great!" There was no mistaking his inner excitement.

"But you've had many other rich experiences in the Arctic. Stu once told me that you and he had taken wolf pups from their parents in the Eureka area of Ellesmere. You've had experiences with muskoxen, you've worked with other shorebirds, you've travelled much of this intriguing land in many of its moods," I said.

"Yes, the Canadian Arctic has been good to me. It's comforting just to know it's here, especially when I'm back there in Kansas." Parmelee was doing what he wanted to be doing, what he would rather do more than anything else in the world.

"Dave, what's the most exciting thing you've seen up here?"

He looked inward for about half a second before replying.

"A tiny aluminum band."

"A tiny aluminum band?"

"That's right. A small, very worn aluminum band on a turnstone's leg."

"Tell me about it, please."

"Well, I had been collecting some scientific specimens up around Eureka on Ellesmere Island and picked up a turnstone with a metal band on its leg. It was not known whether the turnstones nesting there in the Canadian Arctic were of the Old World Race, which winters in southern Europe, Asia, southern Africa, Australia and New Zealand, or the New World Race, which winters from the southern United States to Brazil and Chile. Most people seemed to take it for granted that they were probably the New World form."

"How did you have it figured?"

"I was neutral. You had to be. There was no evidence one way or the other, then. And the band on my bird's leg was so badly worn that I could not make out any of the numbers on it. But with a magnifier I could read, ever so faintly, I thought, 'Br. Mus.', and figured that might mean British Museum. It later turned out that the turnstone had in fact been banded in Britain four years earlier. So some of the turnstones which spend the summer up here have crossed the Atlantic to do so. Knowledge often comes in very small parcels. That was the first record of its kind," he concluded, looking into the distance, as though living it all over again.

"That's fascinating, Dave!"

"Funny thing about it all," said Dave, "is the fact that we got the documentation on the reverse migration the same year."

"You did?"

"Yes, I had banded a nest of turnstone chicks on Ellesmere that year, too. Six weeks later, one of them was taken off the west coast of Scotland. It had been caught in a mist net by a British bird-banding party. They took the number on the band and released the bird again. And two years after that, a second one from the same nest was found dead on a sandy beach in Portugal."

"Whew! Did you ever try to figure out the odds on that happening, Dave?"

"One in many millions, for sure."

Further checking at Nest 1 revealed that the female did not lay her eggs there. Two birds were, however, seen by Dave and me in that vicinity again two days later. On this occasion, the female settled into a different scrape with her posterior elevated towards the zenith as before. The male then approached the scrape and walked around it, rubbing his shoulder against the female as he did so. Then, positioning himself at the rear, he inserted his beak under the female and raised her off the nest. She then walked off with the male a pace or two behind. Then he caught up to her, they rubbed shoulders momentarily and again the male stopped her by pressing his beak against her breast. This time the female half crouched and the male stepped upon her back, and she spiralled for several seconds with the male beating his wings in order to retain balance. Then her tail came up as his dropped down and the cloacae were apposed for seconds.

The male dismounted, the female fluffed her feathers and the male flew off, followed shortly by the female.

However, later checking of the scrape here revealed that no eggs were ever laid in it, either. Nest 1 was no nest.

Was this area, then, merely an area that the sanderlings used only for courting and mating? Was this then a second territory? Second to the territory over which the males had been involved in flight display, presumably advertising for mates? And what next? Did they have still another territory in which they would nest?

Did the sanderlings really return and establish no fewer than three distinct territories? If the practice were true, it was certainly most unusual. Most birds have but a single breeding territory which they defend.

Fortunately, by this time, Parmelee, by dint of long hours of searching and observation, had turned up a sanderling nest with a single egg. He had marked the first egg and, subsequently, the additional three shortly after their having been laid. Should this nest come through without mishap, he was in a position to establish the incubation period for that pair of sanderlings. This is the period of time from the laying of the last egg in the clutch until that egg hatches. Parmelee's nest was located in a clump of brown saxifrage and consisted of a neat circular scrape three to four inches in diameter, lined with a few white elkhorn lichens and dead willow leaves. The greenish blotched eggs were singularly handsome.

The party found very few more sanderling nests over the next week or so. We had even tried dragging a long rope over ridges for hours at a time in the hope of flushing incubating birds. The practice was to have a man at each end of the rope while another followed behind, near the middle of the stretched rope, in order that he might better spot the exact point at which a bird flushed. Success was very limited. Birds on the tundra seemed to be decreasing in number rather than increasing as we had expected. Then we began to lose nests of eggs to Arctic foxes. At first we thought that the marauders might be long-tailed or parasitic jaegers, which were represented in some numbers. However, one day we watched an Arctic fox loping along with his nose in our tracks. He followed them up to a nest that we had just found. He immediately swallowed the eggs and then defecated in the nest before leaving.

Dave Parmelee, very concerned, called a conference. What could we do about it? He reminded us that we had only one sanderling nest, Nest 2, in which it might be possible to establish the incubation period. We had a short discussion and agreed that the only possibility lay in erecting a blind near Nest 2 and manning it for twenty-four hours a day in order to guard it against predation. Dave could not afford to lose Nest 2.

Accordingly, the blind was erected about sixty feet from the nest in question and four of us agreed to take six-hour turns in the blind daily, until the eggs hatched. Besides performing guard duty, we would record every event observed at the nest during our respective vigils. This included feeding periods of the incubating bird, time on and off the nest, altercations with other animals,

Rarely found nest and eggs of the sanderling.

responses to our presence as well as to the presence of other forms, and so on. It was a major undertaking, especially as we all had other interests to pursue. We excused David Gray as he was very heavily committed to his muskoxen study in which he hoped to earn his Ph.D.. My turn was from 2400 to 0600 in the early morning.

Dave had live-trapped and banded each incubating bird and had colour-marked each one on the throat or wingbar with a felt pen of a different colour. This meant that the incubating bird could be identified without flushing it. This was particularly important since alert jaegers were always about and were likely to spot an open nest brought to their attention by our visiting it. Dave had hoped to band both male and female but this was impossible. We became increasingly convinced that only one of the pair assumed the incubation duties .

The reasons for our assumption were straightforward. In the first place, we were not able to obtain any evidence that a second bird shared the incubation of the eggs. We never at any time witnessed an exchange or attempted exchange of the sexes at the nest, as is so common with many other birds. In addition, even though the incubating bird sometimes remained off the nest continuously for periods of up to seven hours or more, never did a mate show up to take over. In the final analysis when all of the records were in, members of the party had checked incubating birds a total of 478 times. The single individuals that Dave had banded and colour-marked were on the nest on every occasion. Some were bright males; others, pale females. Only one bird incubated at a given nest.

What puzzled us more than anything else was the inescapable fact that the pairs split up suddenly after the clutch of eggs had been laid. We were all familiar with the situation in phalaropes, those small shorebirds in which the females laid the eggs but the males assumed all of the household duties thereafter. This included the acceptance of the total responsibility for the incubation of the eggs and for bringing up the chicks. Then there were many shorebirds in which these duties were shared. Clearly, neither of these patterns of behaviour applied to the sanderling. The sexes did not share the incubation duties, nor did only one sex assume them, as in the phalarope. Dave had determined that either male or female assumed all of the duties at the nest; never both.

How was the decision made concerning which sex would take over at the nest? Was it made purely by chance? Did the member of the pair that first returned to the nest at the conclusion of egg-laying settle over the eggs and immediately develop a high level of aggression towards all other birds in relation to protecting the nest? Did such a level of aggression extend even to its own mate?

This theory, involving the sudden development of a high level of aggression on the part of one member of the pair, did not seem altogether illogical in the light of what we had seen regarding aggressiveness in the species. Pairs were intolerant of strangers. However, we all, I feel sure, had niggling reservations about its application in the light of other information gathered. Almost from the outset, at least from the time the males had secured mates following flight displays, the pair bond between the sexes seemed very strong. We saw pairs on the tundra on many occasions and they were inseparable. The courtship was relatively long and elaborate. Whenever one of them flew off, it was immediately followed by the other. How could such a strong pair bond be so suddenly and totally ruptured?

While observing the incubating bird at Nest 2 over the extended period of fourteen days, it became evident that birds nesting in such remote areas had significant problems. Much of the weather appeared very unfavourable with the prevailing cold winds off the Pole and the mercury so often hovering around the freezing point. Snow cover persisted well into summer and could conceivably so shorten the breeding season that the offspring might not have time to develop sufficiently before migration was upon them. Foxes, jaegers and gulls were always on the alert for eggs. As well, herds of muskoxen and caribou often moved across nesting areas and could easily trample the nests. Even wolves and Arctic hare had to be considered, as we were to learn later.

On July 5 I set out from camp to try for some photographs of the attractive Peary caribou with antlers in the velvet. I had noticed three of them feeding on the very sparse year-old vegetation a short distance northeast of camp, and set out down a dry creek bed in an attempt to get close enough for a sequence of their trotting gait. I finally got in range, but they moved off halfheartedly without breaking into their matchless, high-stepping trot. I didn't bother to film them. But as the caribou moved off, a sanderling suddenly took to the air from a slope above them and flew across the narrow valley and around a low hill about two hundred yards away. I felt that the bird might be either returning to its nest from feeding or going out to feed, so quickly decided to follow it to the point at which it had vanished. Here, on the far side of the hill, a gentle slope of gravel dotted with clumps of saxifrage and sparse trailing willow stretched before me. Since it looked like an ideal nesting area for sanderling, I began checking it carefully.

Coming over a very small rise, I came upon a small shallow pond with a sanderling feeding along its margin. I slowly sat down to watch it, but it almost immediately flew off. I tried to follow it with my glasses, but almost lost sight of it down a slope in the direct rays of the sun. But I got the impression that it had dropped to the ground and run a bit before vanishing. I got as careful a line as possible on the spot where I thought it had vanished and then crossed to the slope and began to move slowly down it.

The sanderling flushed when I was still about a hundred yards distant. It seemed to have risen some distance to the right of where I thought I had last seen it. I sat down again and waited. In a few minutes the bird returned again, alighting down the slope. Then it immediately ran fifty or sixty yards across the gravel and vanished. This time, I kept my glasses glued to my eyes while checking rocks, different sized clumps of saxifrage and conspicuous patches of gravel as reference points around the spot where the bird had vanished. I checked again with the glasses, then lowered them to be certain that I could find the reference points with the naked eye when walking forward. I had hardly started down the slope before the bird flushed again. The reference points did the trick and in two or three minutes the scrape with its full complement of four eggs lay before me. This was Nest 6. The site was unusual when compared with the others. The nest was located in a depression between humps of heaved, dried and caked mud.

When I came over the hill two days later to relieve MacDonald in the sanderling blind, I found him throwing things at three wolves. I had a shotgun under my arm in case it became necessary to scare off foxes or jaegers.

"Man, am I glad to see you." Stu breathed as I walked up. "I don't know what I'd have tried next."

"You've got company," I said, glancing at the three wolves, which had moved off thirty or forty feet. One of them had lain down, nose on paws.

"I've had more company besides those three," he said, indicating the wolves with his gloved hand. "Awhile ago, I got so cold and stiff that I decided to do a few pushups in the blind to get the circulation going. When I stood up again, I glanced out of the small window and spotted a black Arctic fox loping down the hill following our scent trail to the nest. I had no time to step out and load the gun so yelled as loud as I could. The fox hesitated and began to circle so I loaded the gun, got out of the blind and dusted him over the hill. I checked the nest because I couldn't be sure that the fox hadn't got the eggs before I spotted him. The nest is fine, thank heaven." Stu heaved a sigh of relief.

"But what about these fellows? When did the wolves show up?"

"Well, I climbed into the blind again, zipped it up to keep out the wind and started to enter the fox episode in my notes. I had just about finished when I heard a sound behind the blind. I thought that you had slipped up to relieve me without my noticing, but you didn't answer when I called out. I unzipped the lower corner of the blind and stuck my head out to see where you were. I looked right up into the eyes of that female wolf that has been tagging along with the others. Her head wasn't more than two feet from mine! Do you know, I thought for a split second that it was a polar bear!"

"Polar bear?"

"I don't know. Maybe I was thinking polar bear. After that one Phil and Dave saw a couple of days ago."

"I can imagine the feeling," I said. "You've had quite a long day in your six hours."

But Stu hadn't finished. "You know that damned wolf just looked me square in the eye for several seconds and snuffled her lips at me. Then she backed off a bit. I jumped out and threw a shotgun shell at her. It missed and went skittering across the gravel. She chased it, picked it up, chewed it a bit, dropped it and then came back. I picked up a handful of gravel and threw it into her face and what did she do? She chased the larger stones bouncing over the tundra and then came back to within six feet of me.

"Then the other two wolves moved in and started jumping around when I threw gravel at them. By now, they were all pretty close to the sanderling nest and I was afraid that one of them would step on the eggs. All of our work on this sanderling would have been for nothing, and it's the only nest on which we have all the data. I tried to work them back behind the blind and away from the nest, thinking, too, that I might get a chance to dodge into the blind for my shotgun. That's when you came up."

"I wonder what they might have done in the end," I said. "I don't like the idea of the others moving in, somehow."

The wolves finally left at a slow trot, moving down the side of the ridge towards the broad flat area below. The female lagged well behind as usual.

"She's a pretty unpredictable character, that one," mused Stu. "And very venturesome. I wonder if she has pups somewhere. She had a lot of fresh blood under her chin today."

I took over in the blind, adding a few more interesting items to my notes during the ensuing six hours. Fortunately, neither the fox nor the wolves returned. Once, the sanderling rose suddenly from the nest and ran, low to the ground, to a point perhaps twenty feet behind the nest, where she squatted motionless beside a clump of saxifrage. Glancing up, I noticed a parasitic jaeger cruising over. After it had passed out of sight, the sanderling scampered quickly back to the nest and settled. On mulling this over, the survival value of such behaviour seemed evident. Had the jaeger happened to land near the nest and flushed the sanderling, the eggs would have been forfeited immediately. Had the predator landed near the crouching bird and flushed her, the eggs would not likely have been spotted. Still later, four sanderlings landed in a group forty or fifty feet from the nest. The incubating bird left the nest and ran at them with wings extended, feathers ruffled and scapulars raised. The four flew off hastily. No doubt about it, the incubating bird was aggressive.

On July 12 I was hiking over to a small pond south and east of Nest 2, where we had been manning the observation blind, when a sanderling flashed by me close to the ground. Not sure whether I had flushed the bird or not, I took up station some distance back, hoping that the bird might reappear either on its way to or from a nest. A very cold wind gradually penetrated my heavy clothing as I remained in position for just over two hours. I stood up and walked to the edge of the small pond in order to view things from there. I immediately heard the low notes of a sanderling and soon noticed it in the shallow water at the margin of the pond. In a moment, the bird flew off in a northeasterly direction, but I lost it due to the tears streaming from my eyes from facing into the cold wind. But I did have some sort of line on the bird's flight path; it seemed about the same as in my first sighting.

Moving to approximately the area where she had vanished from my view, I staked out a sizeable piece of terrain, using rocks and clumps of vegetation as reference points, and spent the next three hours patrolling it foot by foot in the hope of flushing the bird. By now, it was about time for me to hurry back to camp, seize a bite or two and relieve Stu in the blind. I decided to take a route slightly to the right of a direct line to camp, since this was the direction in which the bird had disappeared from view. Looking up, I noted a long-tailed jaeger flying along a ridge a short distance away and parallel to me. I was just crossing a small gravelly gulch about a foot deep when a sanderling flushed almost at my toe and began to run about, mewing agitatedly, with feathers fluffed out. I glanced quickly at the nest before me and immediately became aware of the jaeger, by now hovering directly over the nest. There was little choice. The predator had spotted the eggs and they would be taken the moment I left. I collected her for the National Museum.

The nest, our tenth, was beautiful in the extreme. The neat round scrape was located in the centre of a clump of saxifrage. It was lined entirely with white elkhorn lichens, the four somewhat pointed, greenish, living jewels completely filling the small cavity.

Parmelee had been spending every available hour combing much of an area about sixteen miles in diameter in search of sanderling nests. He seemed tireless in his efforts. I often wondered just how much sleep he actually got. He seemed depressed on July 15 when we were having a quick cup of coffee with some biscuits that I had hastily baked in our cubic-foot tin oven over one of the burners. He had just come out of the blind after his six-hour stint.

"Disappointed today, Dave?" I asked.

"Concerned might be a better way of putting it," he answered, draining his cup. I thought that he looked paler than usual. Tired, undoubtedly.

"Any particular reason?"

"We've got an awful lot riding on Nest 2, you know. Suppose those eggs don't hatch! She was off the nest continuously for seven and a half hours back on the sixth of the month. And today she was away for five and a half hours without returning to warm the eggs once. They must be close to the hatch by now because I've already got chicks on the tundra. You know how cold it has been. I don't think that Phil has recorded an air temperature higher than 42° F. I don't know that the embryos can take that much chilling."

"Only time will tell, I guess, but that's not very comforting," I said. "It can't be long now, surely."

"I've been lucky in the past up here, and I hope she holds," Dave countered. He failed to mention the incredible number of hours of work that went along with the bit of luck.

We spent a series of very anxious hours, beginning on July 17. One of the sanderling eggs in Nest 2 had a hole in its side and was filming over with clotted blood. The hole was not the result of pipping by the enclosed embryo, since the tiny fragments of shell adhering to the membranes were directed inward rather than outward. Which egg was it? Egg No. 4! The one that Dave needed in order to establish the incubation period.

Parmelee was devastated. He felt that he had imposed severely upon all of us in the rigid observation schedules that we had shared.

"And all for nothing," he murmured. "I'm awfully sorry, fellows."

"Hey, just a minute! Look what we have learned about sanderling behaviour at the nest." This from Taylor.

We all wondered how the mishap could have occurred. Then Stu remembered having seen an adult hare with one of its youngsters near the nest on the previous day. He felt that probably the young hare had stepped upon the egg in its play and pierced it with a needle-sharp claw.

Be that as it may, a second hole appeared in the same egg later on the seventeenth. And this perforation was clearly from the inside. The embryo in Egg No. 4 was still alive! The egg was pipped. Maybe we were out of the woods after all. But again, the embryo might not be able to extricate itself from the enclosing membranes and shell.

We were elated that Dave Parmelee was in the blind on July 19 when the critical egg hatched. The sanderling left the nest, carrying a half eggshell in her beak. Parmelee jumped from the blind and found the chick from Egg No. 4 still curled up in the remainder of the shell. He had the time of hatching as accurately as that were possible. He rapidly calculated the incubation period: "Twenty-four days, six hours, fifteen minutes, give or take thirty minutes!" Dave threw his toque into the air in excitement. I'd bet that his smile greatly exceeded that of the Mona Lisa.

Incubation period is established as the time elapsing from the laying of the last egg in a nest until that egg hatches. In most birds, incubation does not begin until all of the clutch has been laid. Many arctic-nesting shorebirds tend to lay their four eggs at intervals somewhat exceeding a day in length, perhaps four eggs in five days. Since incubation begins with the full clutch, eggs laid earlier are exposed to the chill of the tundra until the last is laid. The first egg is chilled the most and the last, the least. Theoretically then, the eggs ought to hatch in reverse order to that in which they were laid. The last one laid should hatch first, since it was least chilled; the first laid, last and the others in order. And this is what actually happened at Parmelee's sanderling nest.

While Dave was engaged in banding the four chicks later, the female flew in and landed beside him. She immediately fluffed out her breast feathers and the youngsters crawled beneath her breast and were brooded, barely inches from Dave's knee.

It was only after another two seasons' work that Parmelee was able to unravel the most significant elements in the roles of the sexes in the breeding behaviour of sanderlings in the Canadian High Arctic. What actually happened when the two sexes separated so completely after the laying of the initial clutch of eggs?

Dave discovered that the answer was simple enough. The female turned that clutch over to the care of the male and then she established another territory and laid a second clutch, which she alone incubated.

Did this behaviour pattern make sense? What was its survival value? It was necessary to recall that sanderlings nest in very remote northern areas where the chances of survival are low, partly because of the vagaries of climate. By the simple expedient of laying eight eggs rather than the usual four, the birds were virtually doubling the chances of survival. And were a predator to take one of the broods, there was always the chance of survival of some of the members of the second brood.

I still have difficulty fully comprehending the evolution of the practice of a single pair of birds occupying four distinct territories for reproductive purposes. An advertising territory, a territory for courtship and copulation, a territory in which the male incubates eggs and still another occupied by the incubating female. I know of no other shorebird that follows this pattern.

But perhaps Dave Parmelee will turn up another. Who knows?

Tundra Wolves

It is hard to know what to make of the wolves of Bathurst Island. Many a tale has been told describing the voracious, cunning, cruel, savage, cowardly nature of this evil, terrifying beast that would kill entire flocks of sheep or herds of deer and cattle only in order to gratify its insatiable lust for blood. In the dead of winter, they have pursued and dispatched sleigh loads of people, together with their unfortunate horses. Since they hunt in vicious packs of up to four hundred members, no animal upon which they wish to prey is safe. Their unprovoked attacks upon humans are legion; the dead and dying have been consumed with the greatest of ferocity. The tales come from Siberia, Spain, Germany, France, Britain, North America.

Shortly after my arrival in camp from Resolute, I noticed a lone bull muskox on the far side of the valley northeast of our camp. MacDonald told me that he had been in that general vicinity for several days, that probably he was an outcast from the herd some three miles away. They had dubbed him Eeyore. We developed the habit of checking the muskox as soon as we rolled out in the mornings. On three consecutive days, he appeared to be in almost exactly the same place. We wondered if he had moved off while we were asleep, only to return later. Stu and I somehow doubted this and decided to check his condition.

We hiked across the valley, coming upon the muskox from behind a narrow ridge that offered a bit of cover. The muskox was lying down quietly on the edge of a partially bare patch of snow and gravel, probably resting, we thought. We took several photographs of him before slowly inching forward. He slowly turned his head in our direction as we approached and attempted to raise the forward part of his body onto his front limbs, but was unsuccessful and settled back again. It was apparent that the muskox was in a bad way, his physical condition critical.

The next to last chapter of Eeyore's life was clearly etched in the gravel and snow around him. Tracks showed that his powers of locomotion had been failing badly; he had moved but thirty-six feet in the preceding two days. Eleven very small piles of dung were closely clustered in the small patch of gravel in which he lay, and in order to arrive at his present position, he had been forced to drag himself forward for the last fifteen feet. When we gently stroked him on the shoulder, he attempted to hook sideways with his great curved horns, but the flesh was weak.

Golden-grey fur covered the hump and middorsal region, the sides were dark brown, the face still darker and the lower legs a bright silver. The lustrous eyes were shades of blue ringed with reddish-umber. We pressed his torso gently. He

A member of the pack, quietly awaiting the kill.

was very thin, the ribs prominent ridges under our fingertips. The eyes were partially closed except when we touched him. He no longer struggled. He appeared to be failing fast.

We watched the old monarch for an hour, standing close beside him. Breathing rate during that hour decreased from forty-seven breaths per minute to eleven. Then breathing became irregular, alternating between slightly deeper inhalations and very shallow, almost imperceptible ones. The flaring nostrils were the only indication that he was breathing at all. Forelimbs began to settle outwards, ever so slightly, with each succeeding breath. The eyelids began to quiver. The end was near. Then a series of the gentlest breaths imaginable. The light went out of the magnificent eyes, the lids closed and breathing ceased altogether. The transition from life to death was barely perceptible. Eeyore slipped away quietly, like a wisp of fog from a valley in calm air.

An aged and helpless prey animal had died peacefully in wolf country? Surely, this could not have happened. The argument that predators cull the herds of game animals and thus remove the sick and infirm, that they show no quarter to a prey animal in distress, has been widely accepted. Yet, for several days, Bloodface, the powerful male wolf, had been occupying a knoll scarcely a quarter of a mile distant. His tracks were on the ridge just above the failing and helpless muskox. Eeyore would have presented no problem to the skillful, powerful carnivore. Yet Bloodface had successfully attacked and killed a healthy bull instead. Why had he bypassed Eeyore? Was he unaware of Eeyore's plight? Did he know that the old gaunt muskox would offer little in the way of food, that the prize would not be worth the effort? Had his lust for blood been temporarily dampened? Was he not typical of his kind? Or were there other reasons, unsuspected by us, why Bloodface had trotted by without confronting the steadily weakening Eeyore?

We encountered wolves on many occasions during our stay on the Island. On my second morning, Phil awakened me from a deep sleep.

"Hey Cy, there's a wolf at the door!"

I pulled on my pants and rushed out to find Bloodface on the carcass of the muskox that he had killed two days before, which Phil and the others had brought back to camp. I was amazed at the power in the wolf's jaws as he proceeded to cut through a rib with a single chomp. He did not back off when I walked to within three or four yards of him. At length, he had apparently satisfied his hunger and trotted slowly across the river to a high knoll, where he curled up and settled with head low. He returned later in the day, but not to feed. Rather, he was interested in getting at David Gray's dog, which he had left in the vestibule of the bunktent. While Bloodface ripped a portion of the nylon wall rather badly in his attempt to dispatch the dog, he was not able to come to grips with him and retreated.

I was reminded of the experiences of Freuchen in Greenland during the winter of 1907-08 when he occupied his isolated station near the great ice cap in order to obtain weather data. Tundra wolves killed all seven of his sled dogs and attacked the dog teams attempting to bring in badly needed supplies for his long and lonely vigil. Apparently, there is often little love lost between wolves and domestic dogs.

I was impressed by the direct way in which Bloodface went about whatever he had in mind. There was no dissembling, no subterfuge, no semblance of cunning. He took the shortest route to the carcass even when we were standing nearby. He trotted directly to the vestibule in which the dog had been left and tried to enter. I had expected the wolves to be wary, to stalk their intended prey, to hunt in a pack, to exhibit cunning and fear of man, to maintain a safe distance at all times. These, I failed to witness. Nor did our wolves kill indiscriminately for blood. Rather, they tended to consume what they killed before bringing down another prey animal.

The total absence of fear of man exhibited by these tundra wolves was unexpected on my part. The remote northern Island was not occupied, even by Eskimo, and it was highly unlikely that any of these carnivores had ever seen people before. I suspect that my problem lay in the fact that I expected them to be wary of strangers, of totally unfamiliar forms of life. I was not on their wavelength. I was thinking like a human, a product of modern industrialized society rather than responding as a wild wolf might.

On reflection, it struck me that what had escaped me was the possibility of two diametrically opposed responses towards unknown forms, suspicion and

Portrait of tundra wolf.

curiosity. The tundra wolves, in their island isolation, had evolved curiosity rather than wariness and distrust. They had certainly not learned to fear man.

Whenever we met the wolves on the open tundra, they never failed to pay us a visit, to inspect us at close range. At first, they circled rather widely, but the circles grew steadily smaller until the wolves were very close, with ears pricked and eyes levelled. They sniffed constantly with muzzles raised as though attempting to comprehend these strange two-legged animals which had suddenly appeared in their environment. They behaved in the same way whenever we saw them in the vicinity of camp. They all came up directly and visited at close quarters for perhaps fifteen or twenty minutes before moving off.

In retrospect, I am unable to recall an instance in which we had cause to fear them. They were always predictable insofar as their relations with us were concerned. On mulling this over at some length, I did not feel that their behaviour towards us was inexplicable. The wolves on Bathurst had undoubtedly lived for generations in the absence of man. They had learned to live off the land, their hunting behaviour directed towards the resident big game animals, the muskoxen and Peary caribou. They, too, were creatures of habit, as were the root-foraging Arctic hares.

It seemed strange that the wolves did not hunt in a pack, taking advantage of their numbers. When Bloodface dispatched the bull muskox, he did it alone. We secured but one record of the wolves pursuing and killing prey in a different manner. This occurred on June 16. The wind had blown a tremendous gale out of the NNW all night, making sleep nearly impossible. I got up early, had a bite to eat and then headed out with a thermos of coffee to relieve Stu in the sanderling blind, as I knew that he would be cold. Stu was very excited. He had just witnessed another wolf kill, this time a caribou rather than a muskox. The strategy employed by the wolves had been different.

The pack of four, consisting of two large males, a large female and a yearling, had approached from the west rather than from the east, as was their usual direction. They had spotted a lone bull caribou, somewhat detached from the main herd on the broad, low expanse of tundra to the west of the sanderling blind. The low area was wet and marshy with standing water here and there, while a dry gravelly ridge overlooked it from the left. The two male wolves, a very large white one and a rangy grey one, moved slowly towards the standing caribou, which did not seem unduly concerned about their presence, and allowed them to get closer than usual. The grey male then swerved in the direction of the higher ground to the left as the white fellow suddenly galvanized into high speed in a direct rush at the caribou.

The caribou bolted with the white wolf some distance behind but running steadily. Both were throwing up clouds of spray as they plunged through marsh and pools of water. The caribou tried to strike for the higher ground on the ridge to the left where firmer footing would have given him an advantage, but the second wolf was already there, running parallel to him. And there the grey male stayed, clearly herding the caribou and preventing him from escaping to dry ground. The white wolf behind was gaining. The caribou looked back and apparently became panic-stricken, for he began to zigzag. Stu said he thought that he had seen a wolf run full-out before, but the white male now put on a tremendous burst of speed, the like of which he had not seen earlier. In no time, it seemed, he was running beside the caribou.

Then he took a great upward leap at the caribou, locked his jaws in the ungulate's throat, jerked its head sideways and then put on the brakes with stiff limbs angled forward. Both predator and prey skidded some distance from their combined momentum. The caribou's head came down, an antler hooked into the wet ground, caught momentarily, and the caribou did a complete flip, coming down upon its back with a great splash. The grey partner rushed in to assist, and in a moment the caribou lay still.

Fifteen hours later, Stu and I had a chance to hike over and examine the kill. The wind was still blowing a gale, but the sun was shining brightly in the west. From a distance, we could make out two wolves at the kill, the big white male that had brought down the caribou and the inquisitive yearling. As we drew nearer, the white male joined the grey female curled up on the near ridge while the yearling left the kill and started walking towards us. The others joined her and all padded forward directly, as though to intercept us. We wondered if they were about to dispute ownership of the caribou carcass, but the pair turned off to the left and the yearling returned to the kill and remained there until we were very close. Then she waded off through a shallow pool to higher ground, her very distended stomach bulging with its load of fresh meat. In a moment the three had lain down together, partway up the ridge where they were joined by the second male.

The kill proved to be a prime bull with well-developed antlers in the velvet.

Peary caribou, a staple item in the wolf's diet.

He had been almost entirely consumed. A small pool of water nearby was stained red with blood, and skid marks told us that this was the place where the caribou had been brought to ground. The stomach had been removed and lay floating on the water, but the animal had been dragged to slightly drier ground for consumption. Most of the ribs had been severed, while the deep wounds in the neck region gave mute evidence of the powerful grip of the wolf after that final lunge. About all that was left in the way of meat were the tenderloins between the vertebral column and the arching ribs.

There were thought-provoking subtleties involved in this event. Again, the prey animal had been brought down by a single wolf, but this time with some assistance by a second member of the pack, which had served in the capacity of a herder, preventing the prey from taking advantage of higher, firmer ground on which it might have been able to escape. There was no danger of injury to the second wolf nor to the other members of the pack. How important was this strategy in the survival of wolves on Bathurst Island? Predators are often seriously injured when attempting to bring down powerful prey armed with horns or antlers and feet armoured with flint-like hooves or claws. Was it generally appreciated by the other wolves that their assistance would not be required, and was this the reason that they lay down and awaited the outcome? Or was this in reality a learning situation for them?

The two male wolves involved had undoubtedly determined that the initial situation was a favourable one, given the caribou's having allowed them to approach unusually close, the expanse of wet marshy ground directly ahead and the possibility of preventing the prey's access to the potential escape terrain. The wolves behaved as though they were aware of the fact that the caribou would be at a disadvantage if forced to flee over the soft sodden area. We had seen the wolves test a number of herds of caribou by approaching them as though to see how close they could get, but only on this occasion were they seen to actually attack. What factors had precipitated the chase and ultimate kill in this case? Had

Eeyore.

nagging hunger alone cast the final die, or did the decision turn upon a combination of hunger and favourable circumstances?

The two males had played two very different roles. Was this a result of a well-developed habit or was some form of communication involved? Certainly the two had performed as an efficient team, the work of each complementing that of the other. While the result had benefitted the pack, only a single member had been put to the risk of injury.

Certainly, the wolves that we encountered on Bathurst Island had not read any of the standard texts on wolf predation.

Neighbours

The wolves, muskoxen, caribou, sanderling and hare had interesting neighbours. Several snowy owls passed through the area, both the smaller, nearly white males and the larger and darker females. Since lemmings were very scarce, few of these owls remained with us for long. While watching a female snowy owl one day with my field glasses, I saw her disgorge a pellet, the indigestible portions of prey that she had recently swallowed. On examining the pellet, I found it was composed of slender bones, fine white fur and a skull with teeth intact, including needle-sharp canines. It was clearly the remains of an ermine or short-tailed weasel.

When I returned to camp for coffee, I tossed the skull to MacDonald. "Hmm," he said. "Short-tailed weasel, eh?"

"Yes, but this seems a long way north for them, don't you think?"

"Not too far," he replied. "I'll bet that I can guess where you found it." He was grinning, a gentle challenge in his voice.

"On a High Arctic island, 140 miles long? I'll take your bet."

"I didn't mean geographically. You found it in a snowy owl pellet!"

"How in the world could you possibly know that?"

"Because I found one in the same place near Mold Bay on Prince Patrick Island away west of here, a few years ago. It has so far proved to be the only record of short-tailed weasels on that High Arctic island."

The skull that I watched the snowy owl disgorge that day was the only record of the same species that we recorded on Bathurst Island that season.

A very few reddish, swift-flying American knots nested locally, as did snow buntings in their brightly contrasting liveries. Purple sandpipers were present, but their nests were extremely difficult to discover. Lapland longspurs were very scarce, as were buff-breasted sandpipers and greater snow geese. Slim dark gyrfalcons went through on pointed wings, as did their smaller cousin, the peregrine. All three of the swift-flying, gull-like jaegers were there, the pomarine with twisted central tail feathers, the long-tailed with its long slender scissor-like ones and the parasitic with central tail feathers pointed but much shorter than the long trailing ones of its long-tailed cousin. We saw several small flocks of white-bellied brant geese from the eastern seaboard, along with numbers of old squaw ducks and a pair of red-throated loons.

When Stu and I returned one day from having checked some ptarmigan nests, Phil Taylor and Dave Parmelee were in the cooktent with steaming cups of hot coffee and some biscuits which we had made in the tin oven. We poured our coffee and spread some honey on a couple of biscuits.

Female king eider, incubating on the open tundra.

"We got a new species today, fellahs," said Phil offhandedly, looking across the valley through the plastic window. From his voice and manner, I somehow got the impression that he was striving to project a casual front.

"And a darned good species, too," Parm added, a grin spreading across his wind-tanned face. His eyes were twinkling.

"What'd you see?" Stu and I asked in the same breath.

"Not so fast, not so fast," answered Parm. "You did pretty well guessing the longspur the other day. Try this one."

"Yeah, you can't miss this one. It's a piece of cake," said Phil. Again, I felt that he was trying to look speculative and serious, but I could see that he was inwardly excited.

"Pintail," offered Stu.

"No, but that's a pretty good possibility," said Parm.

"Baldpate?" I suggested, recalling all of the baldpates on the Anderson Delta.

"No, siree! You fellows are sure slipping!" This from Parm.

"Duck of any kind?" Stu asked.

"I think it's only fair to say no," Phil grinned broadly.

"We've got all the geese that we can expect here, eh?" murmured Stu, frowning.

"I think you're right, there," agreed Parm, still smiling. There was no doubt about it. He had found the new record an exciting one.

"Okay, pectoral sandpiper?" I ventured.

"Hell, no! Look, fellows, this is a real species," explained Parm.

"A shorebird of any kind?" Stu asked.

"No. I've already told you that this is a real species."

Phil had drained his cup and leaned back on the packing case, straining to

*The muskoxen were
vulnerable to
adverse seasons.*

retain his composure. "This is fun," he chortled. "Two arctic types and they don't know what you ought to watch for in the Arctic!"

"Maybe fun for you," said Stu. "But you're just kidding us. You never saw anything!"

"Oh, yes we did," countered Phil. "Right down in the valley there." He swept his arm sideways, indicating the flat marshy area to the south.

"Damn right!" added Parm. "A real species in the real Arctic."

"That's very helpful," I suggested. "You guys are nuts. A pintail's a real species. So are pectoral sandpipers, baldpates and longspurs. What's all this jazz about a real species?"

"A real good species, we said," Parm smiled. "Don't you guys know anything about the real Arctic?" The emphasis was on "real".

"Okay," said Stu. "Crane."

"No."

"An Alcid?"

"No."

"Kittiwake?"

"No."

"Any gull . . . glaucous . . . Thayers . . . ?" pursued Stu.

Parm shook his head in the negative.

"Any passerine?" I asked.

"No, but you're covering a lot of ground there. Holy smokes, I can't figure why you don't try for a real species in the real Arctic!" This was Parm again.

"No duck, no shorebird, no goose, no gull, no passerine," Stu mused. "Are you sure you're not lying?"

"We're not lying!" Phil retorted. "It's a very real species. And as Parm said, you'd better think of the real Arctic." Phil was enjoying himself.

Stu and I were beat, but it was evident that he believed Parm had seen something exciting.

"Give us a clue or two," Stu suggested. "What about its flight pattern?"

"It doesn't have one," Phil laughed.

"Hey, just a minute! Why didn't you tell us that it was a mammal? You said it was a bird. You call that fair?"

"We didn't say that it was a bird. We said that it was a new species. A new species for our records here," Phil corrected.

"Then that's easy," I said, remembering the skull of the ermine. "Short-tailed weasel. But why hold out so long?"

"Wrong again! But the winter colour is the same in both."

Stu turned to Parm, a warm light in his eyes.

"Congratulations, Dave. Your life's ambition realized, eh?"

"You bet!" answered Parm. "And I got a good look at him. We first spotted him in the valley and then he walked up the ridge and disappeared on the other side. He looked as big as an elephant on that open ground."

And that's how the party recorded its first polar bear.

Rock ptarmigan, the most northerly of all the ptarmigan, were very much in evidence from the first day. They are circumpolar in their distribution and occur on all of the High Arctic islands from northern Ellesmere south. They also breed in northern and western British Columbia, Alaska, Yukon, northern Quebec, Labrador and Newfoundland. MacDonald began his work on the breeding behaviour of this species very early, tackling the study with his usual thoroughness, zeal and the insight gathered from his years of working with grouse. Other members of the party assisted him with the business of data gathering whenever possible, especially at the outset, but all of the major work was done by Stu and his assistant, Phil Taylor. When we arrived, both sexes were still in white winter plumage with the usual black outer tail feathers.

The males began to set up territories almost immediately and the courtship phase devoted to the attraction of mates followed quickly. As the season progressed, I was amazed at the many similarities to the events that Mary and I had recorded in connection with our study of the willow ptarmigan on the Anderson Delta. Flight displays, ground displays, the practice of paralleling in rival males, copulation, the behaviour of the male when conducting the female to the feeding area. Even a few of the vocalizations had comparable elements. However, the croaking notes used by the male when communicating with the female on the ground were very different.

On June 14, Phil and I decided to keep close observation of a ptarmigan hen that had turned up, in the hope of gaining more information. The cock joined her for a few minutes and then quickly flew off, leaving her alone. She remained completely motionless but seemingly alert for a full twenty minutes. Then she began to run rapidly forward with body low, in a curious humped attitude. Then she suddenly stopped for several minutes, turning her head slowly from side to side. All at once, she dropped to the ground, squirming and turning this way and

This bull, unable to forage successfully, weakened and died peacefully in wolf country.

that as she pecked at the vegetation around her. She seemed to be in a shallow depression where she at length became still. Could this be her nest?

We waited for about half an hour and then slowly approached her. Save for the blink of an eye, she did not move a muscle as we photographed her from close range. She was a most attractive sight, lying there in her beautifully barred plumage which blended perfectly with the dry vegetation, pebbles and buffy soil about her. We carefully lined up the location of the nest with reference to two whitish stones resting on the tundra and departed slowly and quietly. She had not moved.

When Phil and I returned an hour later, the hen was gone. But where was her nest? It, too, seemed to have disappeared. We walked carefully to one of our markers, examining the ground minutely before slowly placing each foot ahead of us slowly for fear of treading upon the nest. We looked along the imaginary line in the direction of the other marker. Nothing. No nest, no eggs. Could we have been mistaken? Had the hen merely settled on the tundra for a brief rest?

We searched and searched, one of us on either side of the line that we had sighted, careful not to walk directly upon it. A half-hour later, we spotted a very small area where perhaps the vegetative cover had been ever so slightly disturbed. We dropped to our knees. Did she have eggs here and had she covered them on leaving? We stood again, comparing the small area with the surrounding bit of tundra. Did we imagine that there was a difference between this and that? We knelt again and dug the tips of our fingers into a layer of purple saxifrage, lichens and moss. Underneath, two matchless, deep rusty-red eggs with darker markings came into view.

Bits of the same vegetation had been packed securely between the eggs as well. We could hardly credit our senses. An incomplete clutch of rock ptarmigan eggs lay before us! With a bit of luck and much careful observation, Stu should be able to get both the egg-laying intervals and the approximate time of the laying of the final egg, which would enable him, perhaps, to establish the incubation period. We covered the eggs as carefully as possible, checking again and again to be certain that the site had the appearance of the surrounding tundra. Then we left. We had not gone fifty feet before we noticed a long-tailed jaeger cruising overhead. We watched her closely as she hovered over the spot where we had knelt. Had we done our job well enough? In a few moments the jaeger, having found nothing unusual, flew across to the south bank of the stream. We breathed a great sigh of relief.

In another instance, I was present during the laying of an egg. It was a cold raw day with a strong wind out of the northwest. The hen was crouched near the edge of the nest as I approached. The cock joined her briefly, coming to the nest with short steps, head down and tail blown well to one side in the gale. The cock gave a series of croaking notes while walking around the hen and then flew off. She immediately lowered herself onto the nest and flattened with head into the wind and remained motionless for half an hour before switching ends. Twenty minutes later, she stood with tail down for several minutes and then settled again. I believe that this is when the egg was laid. Almost immediately the hen then began arranging the vegetation below her, covering the eggs. Then she began walking forward, a stride at a time, while picking up bits of vegetation and tossing them towards her flanks on either side. Most of this was caught in the wind and blew away. However, she continued her slow forward progress and the motions of head and neck were continued, but she no longer had any material in her beak.

I was struck by the smooth transition between her actions in covering the eggs, and her apparent egg-covering motions at some distance from the nest. To the uninitiated, perhaps even to marauding jaegers and gulls, she might have been engaged in casual feeding behaviour throughout. Had this strategy evolved in connection with misleading flying predators on the lookout for nests and eggs? Should a jaeger spot her, sense that she was indeed in the act of covering eggs and investigate, it would find nothing.

After walking about twenty feet from the nest, the hen suddenly flew off. I checked the sky carefully, failed to spot any jaegers in sight and then quickly checked the nest. A new warm egg had been added, but the eggs had been bared to the sky by the gale-force wind sweeping across the tundra. Still, the eggs were extremely well camouflaged by their inconspicuous background colour, overlain with dark spots and ill-defined blotches. Having found another ptarmigan nest from a preceding season, with its conspicuous pale bleached shells and white membranes, the importance of camouflage in active nests was at once apparent.

The female ptarmigan proved very attentive during their incubation periods,

*Male black-bellied
plover stands guard
near nest.*

allowing us to approach within inches without flushing from the nest. When we
stroked them gently in order to persuade them to leave and allow us to check the
eggs, they merely ran off a few paces and remained standing until we left. Then
incubation was resumed. After the eggs had hatched, the hen led her chicks here
and there over the tundra as they fed eagerly upon bits of vegetation. At first it
was necessary for the hen to brood her charges at close intervals because of the
cold north winds so prevalent in the region. The chicks grew rapidly, assuming
the colouration of the tundra so perfectly that it was impossible to spot them
until they moved.

When Stu and I hiked down the lakeshore towards Bracebridge Inlet on June
27, in search of geese, we counted no fewer than eighty-four king eiders on the
lake, along with a number of old squaw ducks. The drake eiders were
breathtaking with their blue-grey backward-sweeping crowns, greenish-white
cheeks and bright orange bills with lemon-yellow processes. They were busily
courting their wood-brown mates, raising the forward part of the body and
throwing the head backward towards the tail, at the same time serenading them
with their explosive cooing notes. The incomparable, musical, brass
instrument-like chorus of the old squaw drakes provided a rich accompaniment.
Five pairs of purple sandpipers and a pair of red phalarope were feeding in the
grass along the margin of the lake. The scene was a compelling one with the
greenish opalescent water bordered with sheets of alabaster snow intermixed with
brown vegetation. In the backdrop, great expanses of snow still hung from the
bare summits of hills like white silken curtains on a giant stage.

The wind continued relentlessly throughout the following day until just after
midnight, when it calmed down. The temperature had again plummeted to below
the freezing point, but we decided to take advantage of the lull and attempt to
record the flight song of the purple sandpiper. Again a strange and silent Arctic
in the low light except for the widely spaced cooing of the eiders and a note or

two from the drake old squaws. A single sanderling twittered shortly over the marshy area of the lake in front of us. Perhaps the low temperature had dampered courtship activity pending sure signs of imminent spring.

Failing to turn up any nests of purple sandpipers, we had just started back towards camp when we spotted two white wolves on the northeast horizon, along the ridge above the dry bed of the stream. They appeared large and incredibly picturesque against the lead-grey sky with deep snow stretching away on either side. In a moment, a third and smaller wolf came over the horizon and we realized that our pack was back again. The wolves trotted southward for some distance and then paused to raise their muzzles to the silent skies and howl their long and mournful songs, which drifted clearly to us on a freshening breeze. The sound and sight was stirring and elemental, etching the experience indelibly in my memory. This was the true High Arctic.

Would they come nearer or would they, this time, give us a wide berth? The two magnificent white fellows moved slowly to our right and then began padding towards us. A fourth wolf joined them and the three lay down about two hundred yards away, while the smaller, which we took, provisionally, to be a female, started along the ridge towards camp. She changed her mind after a few strides, broke off and began padding directly towards us. We stood silent and still, awaiting developments.

A black-bellied plover flushed in the face of the coming wolf and she vaulted about as though in search of its nest. But then she spotted two old squaw ducks drifting in the shallow water of a pool and drove forward with great leaps in an attempt to catch them. She seemed to enjoy splashing belly-deep in pursuit, but could only raise her head and look after them as they quickly took wing. She resumed her course in our direction and we stood breathless as she stopped about fifty feet away and scrutinized us carefully with head up and level gaze. Then she lowered her head and trotted forward, stopping only when less than a dozen feet away. She raised her muzzle and sniffed audibly while raising and lowering her nose slightly, the better to get our scent. She sidestepped, circled the strange bipeds twice at even closer range, then paused directly in front of us. She calmly spread her hind legs, hunched her back and urinated. Her sex was clear. She sniffed her urine briefly and became excited. She leaped forward, stopped, leaped to the side, stopped and then circled so close that we could see the flecks in her eyes. She had begun to moult, chiefly on her thighs, some of the old fur hanging in tatters and exposing new, shorter, beige-coloured fur below. She suddenly turned tail, splashing through the water to join her fellows as they rose to their feet and all trotted off over a snow-covered ridge.

On the way back, we flushed a red phalarope from a clutch of beautifully marked eggs in a nest located in a clump of moss and sparse grass in a marshy area. We stood back thirty yards or so and watched the male return to the nest. He wandered about a good deal before finally settling.

"I'm going to bring over a blind and try him from close range the first chance I get," I announced. "Gorgeous bird!"

"I think that you'd be wasting your time," Stu advised. "I've tried them

Nest and eggs, black-bellied plover.

before. I set up my blind at a reasonable distance and waited for the phalarope to return. He wandered around for hours and never did settle over the eggs. Completely frustrated, I had to throw in the towel."

"I'd like to give this one a whirl anyway, but thanks, Stu." I was thinking of the experience that Mary and I had had with the tundra swans on the Anderson. They couldn't be photographed with a blind, either, according to our informant.

Back at camp, I checked with Parmelee, who had for years concentrated his attention upon arctic-breeding shorebirds. He related his attempt to photograph a red phalarope at the nest.

"I set up very carefully on one of them with a full clutch of eggs. I wasn't very close but still the bird wandered endlessly. He approached the nest only once, but so briefly that I wasn't able to get a single shot of him. He left in a hurry, deserting the nest entirely. I'd guess that your chances of photographing him at the nest are slimmer than our chances of turning up a long-billed curlew here, a couple of thousand miles north of its range."

The chance that I hoped for came on the following day about midnight, when once again there was a lull in the driving wind. Though we had been up and out for nearly twenty-four hours, the drop in wind velocity spurred us on. Phil and I set out with a blind and cameras to try the phalarope. Though a bank of pearl-grey clouds hung on the western horizon and a low fog lay to the north, the sky above was clear and the sun was shining. Temperatures were down again. At the edge of the marsh, the moss underfoot was a carpet of bright green and red, sparkling with tiny spherical bubbles of escaping oxygen. From the edges of small puddles, delicate fingers of shining ice probed inward, forming intricate patterns.

We sloshed to the nest of the red phalarope. The male flushed swiftly; the nest had not been disturbed. We set up the blind at a considerable distance, with the sun at our backs and full on the nest. We prepared ourselves mentally for a long and probably unsuccessful wait. To our great astonishment, the phalarope returned in a matter of about five minutes, skirted the nest three or four times and then settled over it. He immediately began to pull dry stems of grass over and around him. His colouration was unbelievable. Rich chestnut-red underparts with a soft bloom-like sheen, white cheeks with contrasting dark crown, bright yellow beak with black tip, greyish-brown back with pronounced streaks of beige. The most handsome shorebird that I had ever seen at close quarters!

Phil and I took several photographs and then moved the blind closer and still closer. The bird returned and settled in minutes. We each took several photos and then reluctantly returned to camp at 8:00 a.m. with the clouds beginning to drift over the sun and the wind rising again. Having had a ball, Phil and I had coffee and turned in at 8:45, still very excited about our experience with the "impossible" red phalarope.

By this time the eider ducks were nesting, too. For the most part, the females chose very exposed nesting sites among the purple saxifrage and gravel, rather than against a rock or in a clump of straggling willow. The incubating birds were easily approached so long as one moved slowly. While the eggs were totally lacking in camouflage, the duck drew a cohesive mat of dark down over them before leaving to feed in the lake, which was slowly increasing in size from runoff water. I found it possible to photograph the incubating ducks from inches and thus capture on film every feather line of their beautiful plumage, the well-defined primary and secondary feathers of the wings, the finely patterned breast, the darker lines on head and neck. From directly above, the design formed by the buff-margined upper wing feathers was breathtaking.

I recorded a turnstone on the fourth of July. It was using its stout beak to flip over small pebbles and bits of accumulated vegetation along the lakeshore in order to gather the small food items that it found beneath them, as is the way of all turnstones. I passed the sighting along to Dave Parmelee, our shorebird man, knowing that he would be interested in the record. He told me that he had concentrated upon this very colourful shorebird one season in the Cambridge Bay region of Victoria Island, to the west and south. He had found the experience fascinating. An important part of his work consisted of catching as many turnstone chicks as possible and banding them, in the hope of gaining information about their movements, lifespans, fledging period and the like.

"What was the most interesting part of the study, Parm?" I asked.

"It was all interesting, but the little chicks certainly had me going for awhile."

"In what way?"

"Well, I had banded quite a few of them and then one day I thought that I'd like to have a look at the feeding behaviour of the little guys. So I banded a brood of four young ones and took two of them back to camp and put them in a small cage so that I could watch them more closely."

"What did you feed them?"

"I tried them on different insects, table scraps and bits of meat from my specimens. I got most of my insects from the same place the adults did. From under stones."

"That ought to have made them happy."

Female rock ptarmigan changing in summer plumage.

"Well, it didn't. They wouldn't eat a darn thing!"

"What was wrong?"

"I didn't know. I thought that perhaps they needed a drink, so I placed a small tinfoil dish of water in the cage for them and what do you know? One of them stuck his beak under it, tilted it up and spilled the water over the floor of the cage. I tried again. Same thing! I scratched my head and a tiny light came on."

"Then what?"

"I placed some small strips of raw meat under the plate. They flipped it over in the same way and ate the meat instantly."

"No kidding, Parm?"

"I fed them the same way for several days. I could put food all around them in the cage but they wouldn't touch it unless I put it under the plate or a stone or something."

"So they were prepared to die of starvation in the midst of plenty?"

"You bet they were!"

Both black-bellied plover and parasitic jaeger were incubating pretty steadily when I erected blinds near their nests on July 13. The two nests were only about a hundred yards apart and not far from camp. The problem was that since everyone was so heavily involved with his own specific projects by this time, it looked as though it would be necessary to work alone without a go-awayster to take me to and from the blind. I decided to solve this problem by starting well back with the blind, halfway between the two nests, which would give both pairs of birds a chance to become accustomed to the presence of the new item in their environment.

While the plover did not bother to flush when the blind was put in place, the jaegers gave me no quarter. Both the dark-phased male and the light female flew out to meet me and dived and scolded continuously. They ploughed into my parka hood with force, time after time. I was careful to keep my face down as I struggled to anchor the blind with guy ropes to small boulders. The area was overlain with a good deal of moss, which insulated the permafrost below against the sun. Pegs could be driven into it, but not more than three or four inches at the most.

As it turned out, however, Parmelee was unexpectedly free briefly and took me into the blind to photograph the jaeger. The pair of parasitic jaegers had two characteristically dark eggs and defended them, as we expected, with incredible determination. I decided that predators capable of successfully raiding the nest of a parasitic jaeger would be few and far between. But despite their repeated diving, we finally got the blind up and I focused my camera on the eggs, awaiting the return of the adults.

Within minutes, the light-breasted female landed lightly a few yards behind the nest, lowered her wings and strode forward to the nest. Arriving, she bent her neck down and deftly turned the eggs with her jet-black beak. Then she settled, shimmying from side to side for seconds while assuming a comfortable attitude on the eggs. Her white breast gleamed like fine silk in the low sun. I took several photographs, hoping that they would turn out.

Three days later, one of her eggs hatched. I had not expected the chick to be garbed in such dark, smoky down. I had felt that the adults were pretty demonstrative earlier in the defence of their nest, but as I knelt down to photograph the remaining egg and newly hatched chick, they really elevated the tempo of their attack. Both birds came screaming in repeatedly, striking powerfully at my head and shoulders when they dived, but fortunately my stout parka and hood bore the brunt of their ferocity. I felt lucky to get out alive!

Having moved the blind to a point near the nest of the black-bellied plover, I found myself in a totally different environment. The incubating female flew off quietly, leaving her attractive clutch of greyish, lightly blotched eggs in a shallow cavity. Close examination of the eggs revealed that two of them were already pipped and about to hatch. I resolved to wait long enough to photograph them.

Unexpectedly, the male showed up first, landing about two dozen feet behind the nest with wings raised. The black areas on the underside of the wings were conspicuous until the wings settled at the sides. He made a very attractive sight with the jet-black throat, underneck and underparts contrasting sharply with the snow-white lower belly and the equally bright white area running up the sides of the neck and over the forehead. His brightly checkered back reminded me of that of the common loon.

The much drabber female arrived a few minutes later and ran directly to the nest to settle over her eggs. The male remained in his position for nearly three

Rock ptarmigan incubating in a clump of dryas.

hours, calling to his mate at intervals with a single, piercing, whistled note. She, in turn, never failed to answer in like tone, maintaining the lines of communication with him. But during the last hour, the female had been squirming about on the nest a good deal, often probing beneath her breast with her short beak. Then she suddenly rose and flew quickly off, carrying a very large piece of egg shell. She returned in a moment and repeated the flight as before. After four such trips, she again settled over the nest. I knew that at least some of her eggs had hatched.

At the end of another hour, the female flew off again. She was not carrying anything, but I spotted something wriggling in the nest. I slipped out of the blind and moved forward. Two very young plover chicks, already dried off, nestled on opposite sides of the scrape, an unhatched egg separating them. The pale chicks, speckled in gold, grey and white, were undoubtedly the most attractive shorebird chicks that I had ever seen. I photographed them quickly and left precipitately before the return of the mother to brood her new charges and incubate the remaining egg. Another pair of birds had been successful in hatching a brood in the exacting land well beyond the Arctic Circle.

I felt fortunate, indeed, to witness the occasion. There are those who feel no need for direct contact with the wild and free. The foregoing pages, together with the conundrums therein, are the thoughts and speculations of one who seeks at least a tentative appreciation of the complex nature of the lives of the truly wild.

The Author

Dr. Hampson is a retired Professor Emeritus from the University of Alberta with a B.A. degree in English and Philosophy, and a Ph.D. degree in zoology. He taught at his alma mater for a period of twenty-eight years and has studied and photographed wildlife in many parts of the world over the past fifty years, including East Africa, the Arctic coast, the High Arctic, Mexico, Scotland, the Outer Hebrides and most of Canada. His lectures, radio broadcasts and telecasts run into the many hundreds.

He served as Chief Associate Editor and Photographic Editor for, *Alberta, A Natural History*, which has gone into its eighth printing, breaking every record in Canada for a book in its class. His *Into the Woods Beyond*, was published by Macmillan, 1971.